Fire Stick

A Step-by-Step Guide and Quick Tips for Getting the Most out of Your Fire Stick with Alexa Voice Remote

Nathan George

Fire Stick: A Step-by-Step Guide and Quick Tips for Getting the Most out of Your Fire Stick with Alexa Voice Remote

Disclaimer: Every effort has been made in the preparation of this book to ensure the accuracy of the information presented. However, the information contained in this book is sold without warranty, either express or implied. Neither the author nor publishers will be held liable for any damages caused or alleged to have been caused directly or indirectly by this book.

Published 2022.

Published by GTech Publishing.

ISBN: 978-1-9162113-7-7

CONTENTS

Introduction

The Amazon Fire TV Stick is a revolutionary way of delivering home entertainment. With this wireless device, you can access a catalog of movies, TV shows, documentaries, music, and games you can stream from remote servers on the Internet. You do not need to download any programs, record any programs, or worry about program times. You get to watch your TV programs on-demand.

This guide is to get you up and running quickly with your Fire TV Stick (or Fire TV Stick 4K) with Alexa Voice Remote. *Fire Stick* will show you how to set up your device and the best ways to find movies and TV shows to watch from the large catalog available. You also get to learn how to navigate the myriad of options and several tips and shortcuts that are not instantly obvious from just using the product.

Fire Stick will also cover how to use the Prime Instant Video website on your computer to find and select videos you can add to your Watchlist to make them available for easy access on your Fire TV Stick.

Who Is This Book For?

The Fire TV Stick is a pretty intuitive product, and for some people, the user interface will be self-explanatory. However, some tasks do require step-by-step instructions to perform.

Just to be clear, some of the information you'll find in this book can be found from Amazon's online guides for Fire TV if you carry out a search. However, the information is spread across several web pages and is not available as a single download. If you're comfortable with new tech devices and you're happy with searching for help online, **then this book is not for you**.

This book is for you if you were perhaps disappointed that the product only came with a quick setup guide (instead of a comprehensive user guide), and you want a self-contained guide you can hold in your hand as you use the device. This book has done all the heavy lifting for you in terms of bringing the information together into a single source and adding information not available from the official guides. The information has also been delivered in a user-friendly format compared to the cryptic text in the online help guides.

What You'll Get from This Book

The topics have been organized so that the information you're looking for will be easier to find. With this book, you get a comprehensive, self-contained, guide and you would not need another reference manual. It covers all the product's main functions, and for any regular user, this is all you would need to enjoy a seamless experience of using your Fire TV Stick.

You can use *Fire Stick* as a step-by-step guide during initially setting up your Fire TV Stick and as a reference guide that you can come back to from time to time.

Chapter 1: Initial Setup

This chapter will cover the setup process to get you up and running with your device. The setup process is very intuitive, and if you follow the on-screen instructions to set up your device, you can't go wrong unless there is a technical issue with your setup, and we'll cover these later.

Fire TV Stick Versions

There are currently four versions of the Fire TV stick:

- Fire TV Stick Lite
- Fire TV Stick
- Fire TV Stick 4K
- Fire TV Stick 4K Max

Each device has slightly different features and specifications. Generally, the devices have increasingly more features as you progress through the versions from the Fire TV Stick Lite to the Fire TV Stick 4K Max.

For a detailed breakdown of the differences in specifications between these versions of the Fire TV Stick, visit the product page for the Fire TV Stick on the Amazon store for your country.

Note: There is also a Fire TV device called the Fire TV Cube, which is outside the scope of this book.

This guide covers all four versions of the Fire TV stick. I will be using the general name Fire TV Stick to refer to all the device versions, apart from instances where differences in features and setup are pointed out.

What's in the Box

All versions of the Fire TV Stick have a similar-looking HDMI connector and a micro-USB port for the power supply. The HDMI connector can plug directly into your TV's HDMI port. However, you should use the included HDMI extender cable for the best results.

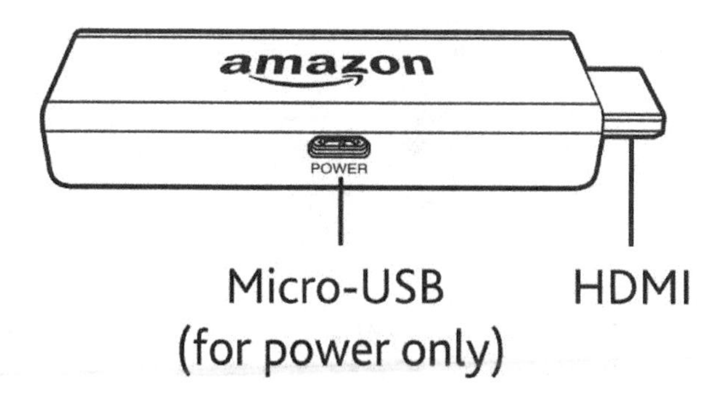

HDMI Extender Cable

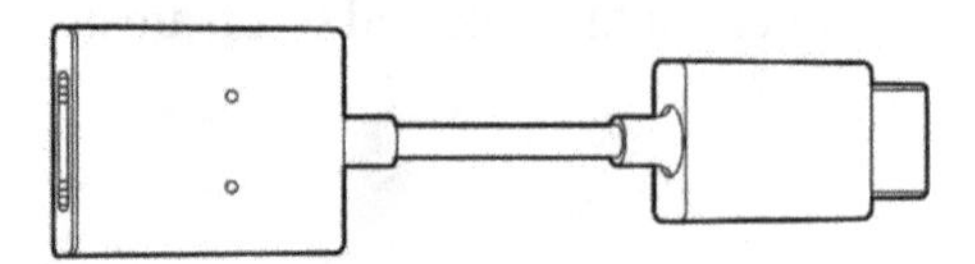

The casing of the Fire TV Stick is bulkier than a normal HDMI cable and may not fit into the HDMI port on the back of some TVs. This is where the HDMI Extender can help. The casing of the HDMI extender is the same size as that of an HDMI cable and will fit any space reserved for an HDMI cable. Also, the HDMI Extender can help with better Wi-Fi reception for the Fire TV Stick.

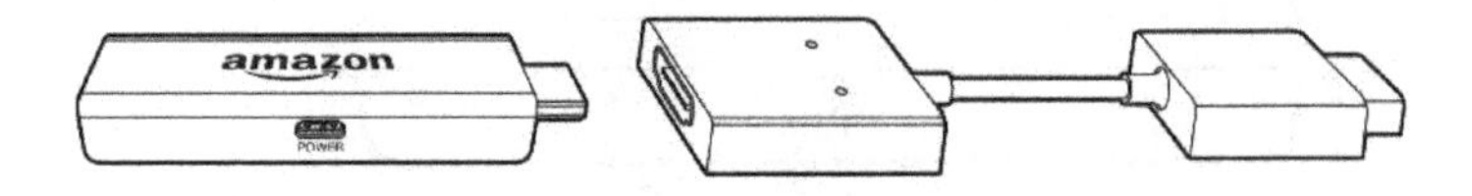

Plug the Fire TV Stick into the HDMI extender, then connect the HDMI extender into an available HDMI port on your TV.

USB Cable

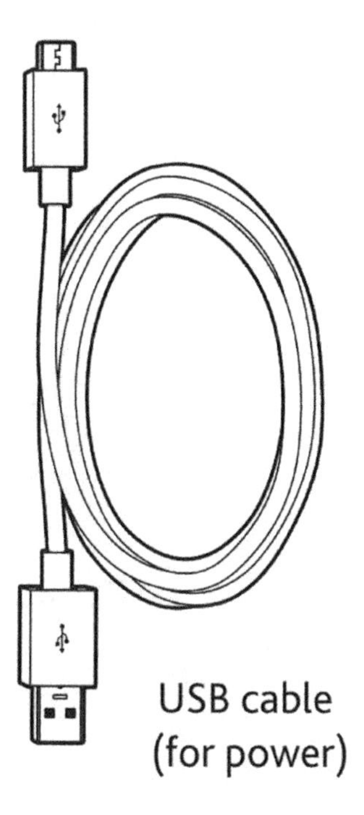

The USB cable connects your Fire TV Stick to a power source. Plug the end with the small connector into the micro-USB port on your Fire TV Stick. Plug the end with the large connector into the included power adapter, which needs to go into a power outlet.

Note: You can also plug the USB cable into a USB port on your TV to power your stick. But this is not recommended as some USB ports do not produce enough power to operate the stick optimally. If you use the USB port for power, the Fire TV Stick may work but not perform optimally, and it would seem like the device is faulty or of low quality. To make sure you're powering the device optimally, connect it to a power outlet.

Also, note that you cannot connect the USB cable to a computer or other devices. The Fire TV Stick only works on a TV with an HDMI port.

Power Adapter

The connectors for the power adapter you get will be based on your region. Different regions have different connectors.

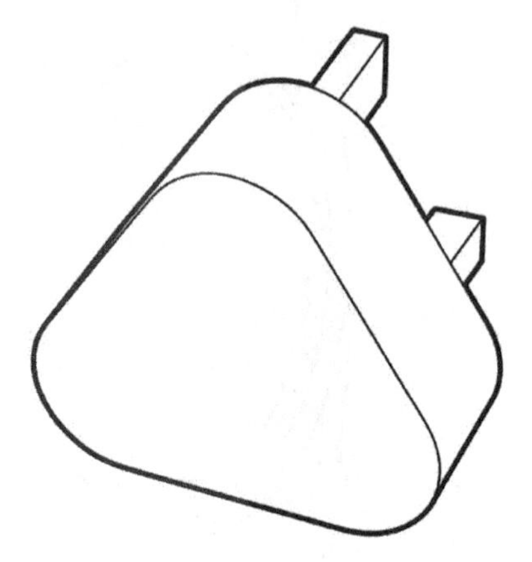

Power adaptor

Remote Control

Three versions of the Fire TV Stick share the same type of remote control.

Device	Remote control
Fire TV Stick Lite	Alexa Voice Remote Lite
Fire TV Stick	Alexa Voice Remote 3rd Gen
Fire TV Stick 4K	Alexa Voice Remote 3rd Gen
Fire TV Stick 4K Max	Alexa Voice Remote 3rd Gen

The remote control is voice-enabled, allowing you to use the Alexa feature that is built into this generation of the Fire TV Stick.

Alexa Voice Remote for the Fire TV Stick Lite.

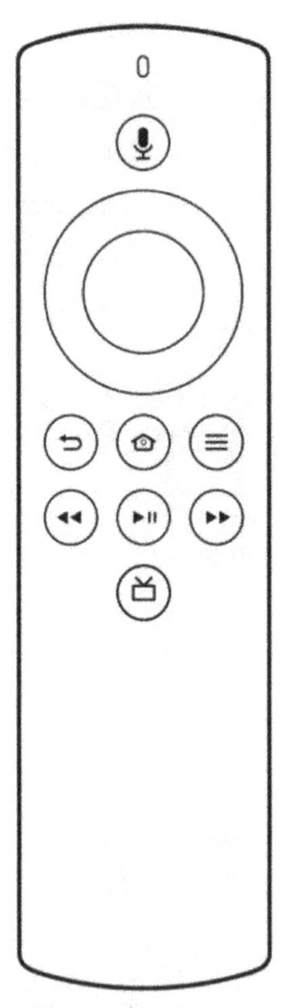

Alexa Voice Remote
Lite

The Alexa Voice Remote for the Fire TV Stick, Fire TV Stick 4K, and Fire TV Stick 4K Max.

This remote comes with extra buttons, including power and volume buttons.

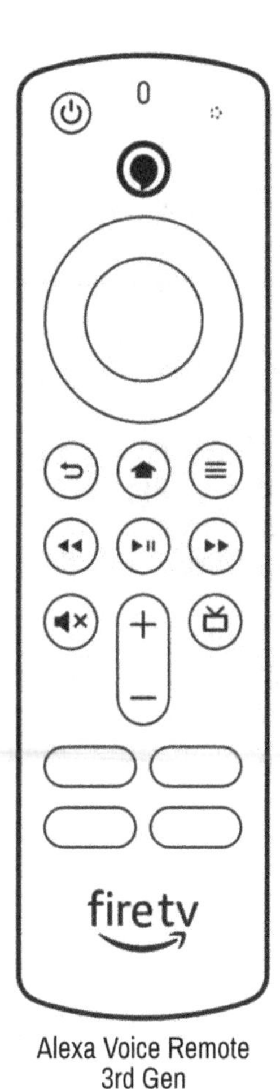

Alexa Voice Remote
3rd Gen

Setting Up Your Fire TV Stick

In this chapter, we will go through the steps to connect your device to your TV and then set up your Fire TV for the first time. This will include connecting to your Wi-Fi network, registering your device to your Amazon account (if not purchased directly from an Amazon store), and configuring the Alexa Remote Control power and volume buttons.

A Short Note About 4K Viewing

You can use all versions of the Fire TV Stick with 720p, 1080p, and 4K TVs. However, a stick will stream content based on the highest resolution possible from the combination of your TV and the stick.

For instance, if you have a 4K version of the Fire TV Stick and a 4K TV that supports HDCP (High-bandwidth Digital Content Protection), you'll be able to stream content at 4K if the content is 4K. On the other hand, if your TV is an HD TV, you'll still be able to use a 4K Fire TV Stick, but you can only stream content at 1080p.

Note that not all TVs advertised as "4K" come with an HDMI HDCP 2.2 port. To get a true 4K input, you'll need to have an HDCP 2.2 port or equivalent on your TV. If your 4K TV supports HDR (High Dynamic Range), then it should have an HDMI HDCP 2.2 port.

The setup process described below covers all versions of the Fire TV Stick. Any area with a difference in setup is pointed out.

Connect Your Device

Connect the Fire Stick to your TV with the following steps:

1. Plug the Fire TV Stick into the HDMI extender.

   Even if the Fire TV Stick can plug directly into your TV's HDMI port, for best results, use the included HDMI extender.

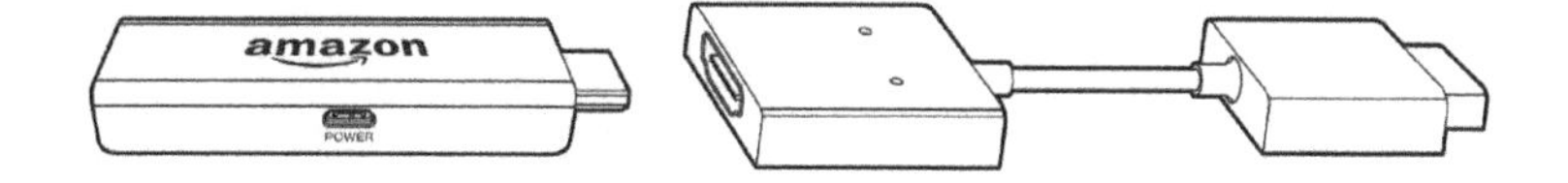

2. Plug the micro-USB end of the provided USB cable into the micro-USB port of your Fire TV Stick.

3. Plug the Fire TV Stick (plus HDMI extender) into the HDMI port on your TV.

   Note: If you have a **Fire TV Stick 4K** or **Fire TV Stick 4K Max** and a 4K TV, connect the stick to a port on your TV labeled **HDMI HDCP 2.2** or **HDMI 4K (@ 60Hz.** This is a 4K Ultra HD port, which is different from a standard HDMI port.

4. Plug the other end of the USB cable into your power adapter.

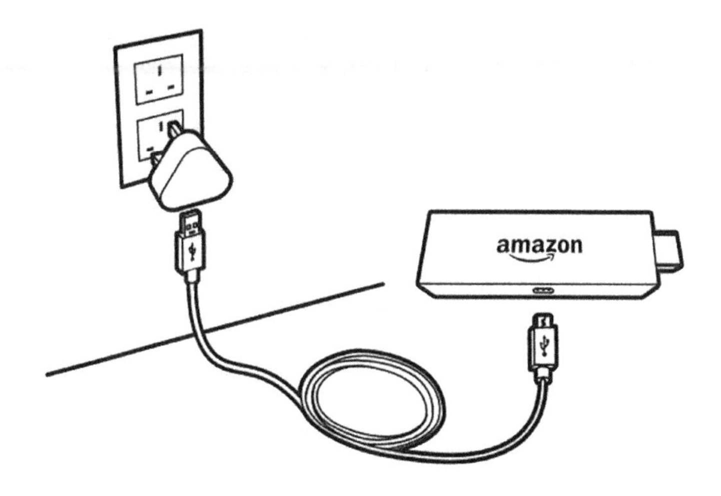

5. Connect the power adapter to a power outlet.

6. Insert the two AAA batteries into the back of the remote control.

Follow the On-Screen Instructions to Setup Your Device

Pairing your remote control

Once you have powered on the Fire TV Stick, the Alexa Voice Remote should automatically be paired with your Fire TV Stick. If the remote is not working, remove and re-insert the batteries. A notification will appear on the screen that the remote has successfully paired.

If the remote is not automatically paired after you remove and re-insert the batteries, press and hold the **Home** button for at least 10 seconds (or until the remote LED begins flashing amber). Your remote should now be paired.

Configure power and volume controls

The Fire TV Stick, Fire TV Stick 4K, Fire TV Stick 4K Max come with remotes that have volume and power controls.

After paring your remote control, Fire TV will display on-screen instructions for controlling your TV's power and volume if your TV is compatible. Follow the on-screen instructions by increasing the volume of your TV using the volume buttons on the remote control.

If an attempt doesn't work, select **No** at the end, and try again. Each iteration uses a slightly different configuration to try to pair your TV with the remote. Eventually, Fire TV will find a configuration that works with your TV. When your remote can be used to power your TV on/off as well as increase/decrease the sound, select **Yes** (when asked if it worked) to complete the initial setup.

Note: If you can't get the volume controls to work during the initial setup, don't worry, you can always try again from the Settings menu on the Home screen at a later point. See **Configuring the Power and Volume Controls** in chapter 2 of this guide.

Connect to your Wi-Fi

Next, connect to your Wi-Fi. Select your Wi-Fi from the list of available Wi-Fi that the device will pick up, then enter your password.

Notes:

- If your network is hidden, scroll down the list and select the *Join Other Network* option, then manually enter your network details using the on-screen keyboard.
- Most Wi-Fi routers now have at least two bands, 2.4 GHz and 5 GHz. For best Wi-Fi performance, use the 5 GHz band on your wireless router if available. For 4K viewing, it is important to use the 5 GHz frequency.
- To use Wi-Fi 6 connectivity with the Fire TV Stick 4K Max, you need to have a Wi-Fi 6 router, which is different from a standard 2.4/5 GHz router. However, the device will work with a standard Wi-Fi router if you don't have a Wi-Fi 6 router.
- Ensure your wireless router or Fire TV Stick are not in a cabinet as it may affect the Wi-Fi signal strength.

Register your device

Next, register your device to your Amazon account. If you purchased your Fire TV Stick directly from Amazon, it should have been registered to your Amazon account already.

If the device is not registered to your account or you purchased your stick from a third party, like a high street store, then you would need to register the device by following the on-screen instructions.

The setup process now provides two options you can use to register your device:

- You can use another device (like a PC or tablet) to navigate to **https://www.amazon.com/code** and enter the code provided on-screen to register your device automatically. You may prefer using this method if you find the on-screen keyboard difficult to use.

- You can manually enter your email address and password to register the device using the on-screen keyboard.

Sleep Mode

The Fire TV Stick triggers a screensaver after 5 minutes of inactivity (this is the default setting which you can change in **Settings > Display & Audio > Screensaver**). After 20 minutes of inactivity, the device goes into sleep mode and a blank screen. So, when you have finished viewing, you can simply turn off your TV, and your device will go to sleep after 20 minutes. You can also manually put the Fire TV Stick to sleep if you want.

Note: To access the **Settings** menu on Fire TV, select the gear icon on the main menu.

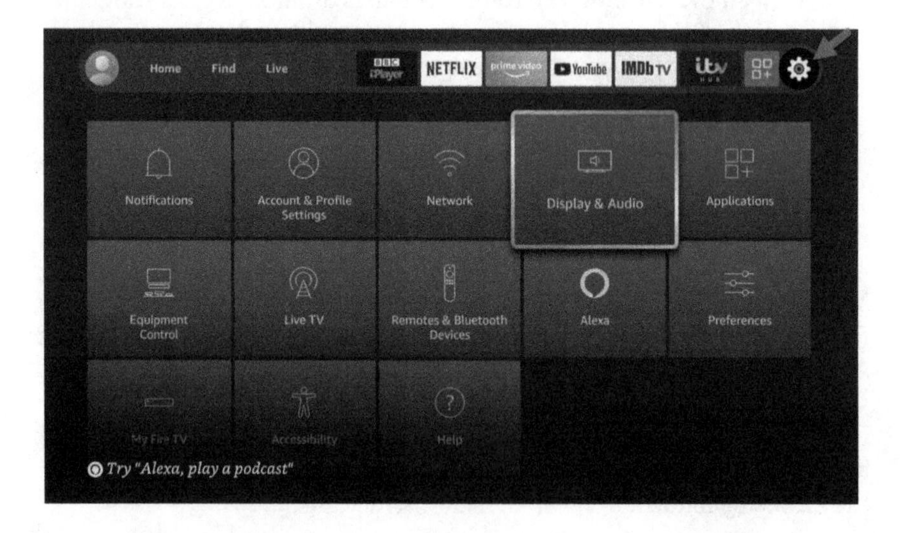

To manually put the Fire TV Stick to sleep, do the following:

1. Press and hold down the **Home** button to open the Quick Access menu.
2. Select **Sleep** to put the Fire TV Stick instantly to sleep.

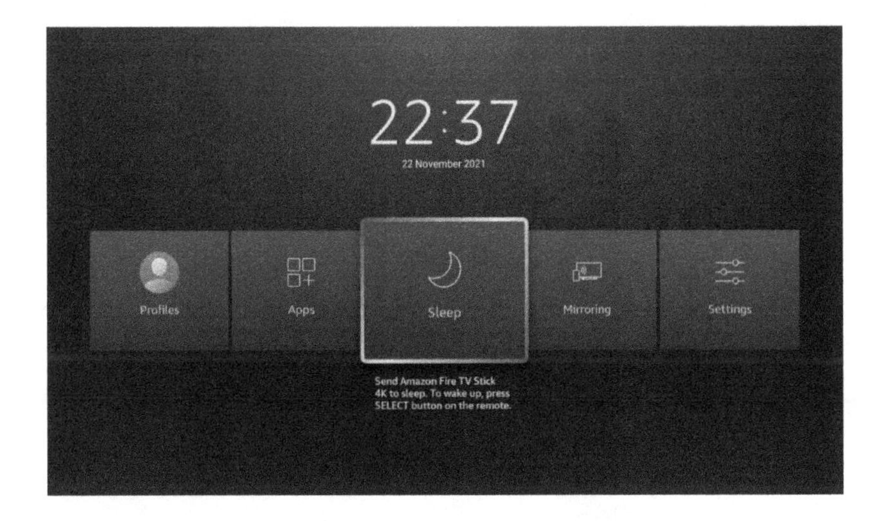

The device itself has no moving parts, so it consumes virtually no power when in sleep mode. Hence, there is no need to turn it off at the power source. Another reason you don't want to unplug the device at the power source is that while in sleep mode, it can still receive important automatic software updates.

To **wake up** your Fire TV Stick, just press any button on the remote control.

Restarting Fire TV

Occasionally you may need to restart your Fire TV due to a software problem. Maybe the screen has frozen, or it is behaving erratically. In situations like this, the best thing to do is to simply restart the device. Restarting your fire stick is equivalent to rebooting the device. Note that **restarting** Fire TV is different from *resetting* it to factory defaults. It simply means turning the software off and back on again.

There are three ways you can restart your Fire Stick:

- **Method 1:** Press and hold the **Select** button and the **Play** button at the same time until the Fire TV Stick restarts. It takes about five seconds.

- **Method 2:** Go to **Settings** > **My Fire TV** and select **Restart.**

- **Method 3:** Unplug the power adapter from the power outlet and plug it back in. This method will be your only option if your remote control is unresponsive as well.

Help Videos in Your Fire TV

There are several useful help videos on Fire TV. You can access them by navigating to:

Setting > Help > Help Videos.

The topics currently covered include:

- Welcome to Fire TV
- How to search for content
- Downloading your favorite app
- Personalizing your entertainment experience with profiles
- Fire TV basics
- Setting up your Fire TV
- Creating, editing, and deleting Fire TV profiles
- Mirroring your mobile device on Fire TV
- Customizing the apps on your Fire TV
- Troubleshooting Wi-Fi connection issues
- Pairing your remote using Fire TV Settings or the Fire TV app
- Fixing a blank screen on the Fire TV device

These videos are regularly updated and would come in handy if you want to familiarise yourself with a feature of Fire TV or troubleshoot a problem.

Chapter 2: Understanding Your Remote Control

In this section, we will cover the Alexa Voice Remote Lite and the Alexa Voice Remote 3rd Gen. You use the remote to access the main menu and your movies, TV shows, games, and apps.

The Alexa Voice Remote 3rd Gen has some extra buttons that are not on the Alexa Voice Remote Lite.

Alexa Voice Remote Lite

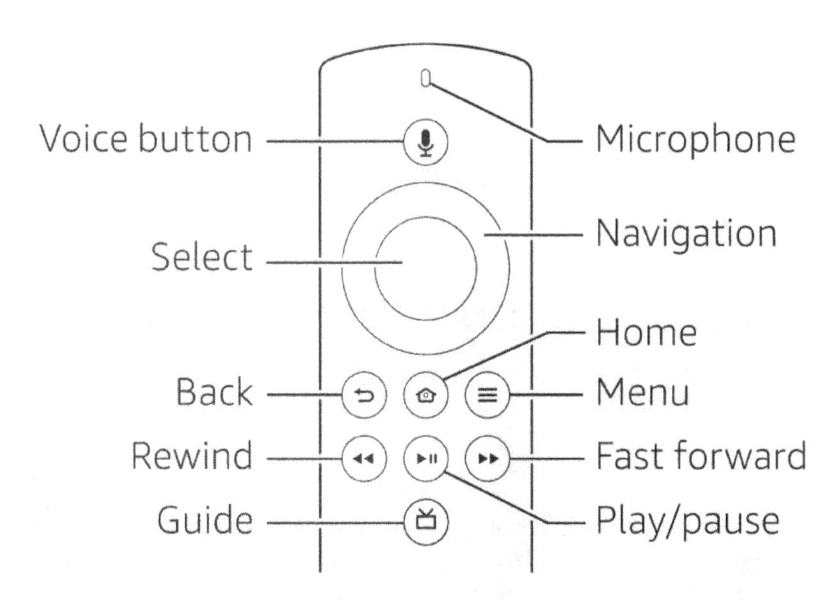

Note: For the rest of this book, I will be referring to the buttons on the remote control by their names. You can refer to this page at any point for the symbol associated with a name.

Voice Button

You press and hold the Voice Button on your remote control and say what you want Alexa to do. Ensure you speak close to the **microphone** on the remote so that Alexa properly hears what you're saying. You can search for movies, TV shows, check the weather, check traffic information in your area, launch apps like Netflix, and many more. Alexa then responds by narrating/displaying the requested information or carrying out the action.

Navigation

You press the 5-way directional trackpad to move up, down, left, or right. You press the middle Select button to select an item, function, or category.

You use the circular trackpad for navigation:

1. Press the top of the circle to **move up** and press the bottom to **move down.**
2. Press **right** to move to the right on your TV screen. Moving to the right from the Main Menu allows you to access the content libraries like movies, TV, games, apps, photos, and more.
3. Press **left** to move to the left on your TV screen. Moving to the left allows you to return to the Main Menu from any content library.

Select

The center of the trackpad has the **Select** button, which you press to select an item, function, or category on the menu.

Home

The Home button returns you to the main menu from any screen on your Fire TV.

To access various features with the Quick Access Menu, press and hold the Home button on your remote. For example, you can use the Quick Access Menu to put the Fire TV Stick to sleep.

Menu

The Menu button displays various viewing options when watching a movie. For example, it enables you to turn subtitles on or off, add items to your watchlist while browsing, remove items from your watchlist, and display various details about the movie or TV show.

Back

This button takes you back to the previous screen or action.

Rewind, Play/Pause, Fast Forward

These media control buttons let you rewind, play, pause, and fast-forward video and music.

Press the Rewind or Fast Forward button once to skip 10 seconds backward or forward.

Press and hold the Rewind or Fast Forward button for a few seconds to make the rewind or fast forward continuous. Once activated, you get a small pop-up screen at the bottom of your TV screen showing your progress and enabling you to cycle through 3 speeds.

Guide

The Guide button (with a TV icon) opens the built-in channel guide on Fire TV based on the apps you've installed with a live programming schedule.

Alexa Voice Remote 3rd Gen

Compared to the Alexa Voice Remote Lite, the Alexa Voice Remote 3rd Gen has some additional buttons for the extra features on the Fire TV Stick versions for which it is used. These extra buttons are described below.

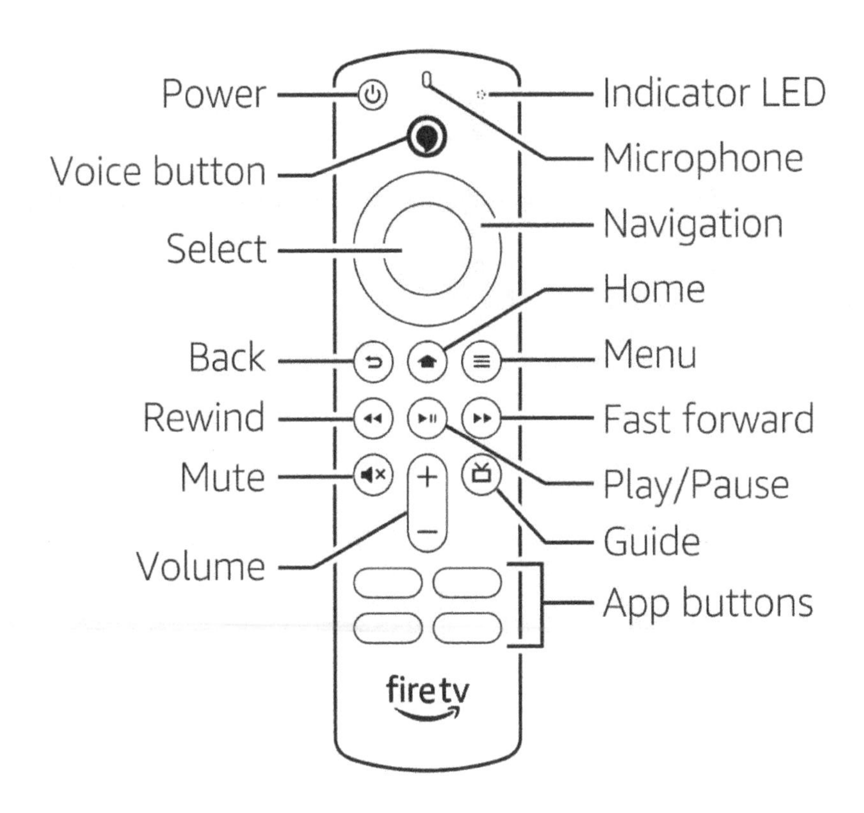

Power button

Press the power button to toggle your TV on and off.

Volume Up/Volume Down

Once your remote has been configured to work with your TV, the volume button can be used to increase or decrease the volume of your TV. With this, you no longer need to reach for your TV remote when you want to adjust the volume of the TV while watching a video on your Fire TV Stick.

Mute

Press the mute button to toggle mute on and off on your TV. Depending on your TV make, if mute is active, there'll be an indicator on your TV telling you that the sound has been muted (usually at the top-right of the TV).

App Buttons

There are four app buttons on the remote. When you press one of these buttons, it immediately opens the associated app.

These buttons are different in different regions. For example, in the US, the app buttons are for Prime Video, Netflix, Disney+, and Hulu. In the UK, Canada, and several other countries, the fourth button is for Amazon Music. In India, the third button on the remote is labeled "Apps", and it opens the screen for all apps when you press it.

Adding an Additional Remote Control

You can pair up to seven remotes or controllers to your Fire TV Stick at the same time. The remote that comes with your Fire TV Stick will automatically pair after you insert the included batteries during setup. To pair an additional remote, follow the steps below.

Note: Before pairing the new remote, ensure you have the old remote at hand as you'll be using it to navigate through **Settings** on your Fire TV to perform the pairing.

1. Insert the batteries in the new remote.

2. On the **new** remote, press and hold the Home button for 10 seconds. The indicator on the remote will slowly flash amber for about 10 seconds before it begins to flash faster, indicating the remote is ready to connect.

3. With the **old** remote, press the Home button, and then navigate to **Settings** > **Remotes & Bluetooth Devices** > **Amazon Fire TV Remotes** > **Add New Remote**.

4. Fire TV will search for available remotes with which it can connect and display them in a list.

5. Using the **old** remote, select the **new** remote from the list. You'll see a message on the screen telling you it has been successfully paired.

You can now use the new remote control with your Fire TV Stick as an additional remote control.

Configuring the Power and Volume Controls

To pair your TV with the power and volume controls on your Alexa Voice Remote (3rd generation), carry out the following steps:

1. On your remote, press the **Home** button, then navigate to **Settings** > **Equipment Control** > **Manage Equipment** > **TV** > **Change TV**.

2. On the **Change TV** screen, select **Change TV**.

3. Fire TV will detect the make of your TV and will ask if it is correct. Select **Yes**.

4. Follow the on-screen instructions to pair your TV with the remote's power and volume controls. Fire TV will give you six attempts, so don't worry if an attempt doesn't work. If an attempt fails, simply select **No** when asked if it worked. Fire TV will take you through the process again with slightly different configurations.

5. Select **Yes** to complete the process when you can use the remote to power your TV on/off as well as increase/decrease the volume.

Your Alexa Remote Control can now be used to power your TV on/off and increase/decrease the volume. This will save you from having to reach for the TV's remote control when you want to change the volume while watching Fire TV.

Troubleshooting Remote Control Pairing Issues

If your remote is not pairing, try the following suggestions to connect your remote:

1. Remove and re-insert the batteries in your remote. Or change the batteries if they're old.
2. Move closer to your TV when pairing the remote. It is recommended that you stay within 10 feet (3 m) of the Fire TV Stick.
3. Unplug your Fire TV Stick from the power source, or adapter then plug it back in.
4. Use the Fire TV app on your mobile phone to pair your remote.
5. If you already have seven remotes/controllers/Bluetooth devices paired, remove one of them before trying to pair the remote.

Chapter 3: How You Use Your Fire TV Stick

There are several ways you can use your Fire TV Stick to stream videos. In this chapter, we'll cover some of the most popular ways you can use the device.

Amazon Prime Membership

The Amazon Prime membership allows you to watch Prime videos as part of the package, including other benefits like music, Kindle books, and quick deliveries for physical products bought on Amazon.

At the time of this writing, Amazon Prime membership fees are:

- $119 per year ($12.99 per month) in the US.
- £79 per year (£7.99 per month) in the UK.

Prime Video Only

You can get the standalone Prime Video subscription just for Prime videos. This is a monthly subscription that you can cancel at any time. At the time of this writing, the price is $8.99 per month in the USA and £5.99 per month in the UK. The good thing about this Prime subscription is that, unlike Netflix, if you cancel halfway through the month, you get refunded for the unused period.

Netflix

You can use your Fire TV Stick solely for your Netflix viewing. The Netflix app on Fire TV is very good and responsive and certainly better than the computer or web-based app. Netflix has a comprehensive list of movies, TV shows, and documentaries. For many people, this can be enough for their online streaming needs.

IMDb TV

IMDb TV is a free streaming service that allows you to stream premium movies and TV shows without needing to buy a subscription. The only catch is that IMDb TV is ad-supported, so you will have ad breaks. However, the ads are well-spaced and do not lessen your viewing experience. You can watch IMDb TV content from one of the featured lists on your home page or by installing the IMDb TV app from the App Library.

Hulu

Hulu is a subscription-based service that allows you to stream movies, TV shows, and live TV. At the time of this writing, Hulu is only available in the US. If you have the Fire TV Stick 4K or Fire TV Stick 4K Max in the US, you can watch Hulu on your device. The fourth app button on the Fire TV Stick 4K launches the Hulu app. You can also watch Hulu by installing the Hulu app from the App Library.

Disney+

Disney+ is a subscription-based service that allows you to stream movies, TV shows, and live TV. In some regions, the third app button on the Fire TV Stick 4K and Fire TV Stick 4K Max remote launches the Disney+ app. You can also watch Disney+ by installing the app from the App Library on Fire TV.

YouTube

At the time of this writing, the YouTube app is available on Fire TV again. YouTube is increasingly becoming one of the top streaming platforms for many people for video content like news, documentaries, vlogs, music videos, and so on. So, having a user-friendly YouTube app on your big screen TV comes in handy as a viable home entertainment option. You can also use the YouTube app to watch videos you've purchased or rented from the Google Play store (external to Amazon).

Add-on Subscriptions

Amazon provides the opportunity to subscribe to some other networks like Showtime, HBO, Starz, and CBS as **add-on** subscriptions. But you must be subscribed to Prime first before you can add these subscriptions to increase your viewing options. Add-on subscriptions are usually on a monthly rolling contract, and you can cancel at any time.

You could purchase videos offered by add-on networks directly from Amazon video. However, it might work out cheaper to subscribe to the network as an add-on subscription if you want to watch a lot of videos offered by that network. For example, if you want to watch all seasons of *Game of Thrones*, it will work out cheaper to subscribe to the HBO add-on channel for a short period (if you are in the USA) instead of buying the individual seasons directly from Amazon video.

Free Streaming Services

There are countless streaming apps you can download to watch free on-demand television content from terrestrial channels like BBC iPlayer, All4, My5, ITV, etc. The apps you get will vary depending on your location. For example, viewers in the US will have a different set of apps compared to viewers in the UK and Europe.

Install Third-Party Streaming Apps Like Kodi

You can install a video streaming application like Kodi on your Fire TV Stick. Kodi allows you to install third-party plugins that may provide access to movies, TV shows, and music that are freely available on the websites of content providers. In saying that, this guide does not endorse the streaming of illegal or pirated content. Installing Kodi on your Fire TV Stick is outside the scope of this book, but there are many sites online that will give you detailed instructions for how to install it. Just type in "*Install Kodi on Fire TV Stick*" in your Internet search engine.

Chapter 4: Amazon Prime Overview

To get the full benefit of your Fire TV Stick, I recommend that you subscribe to Amazon Prime (or Prime Videos) if you're not subscribed already. You could try out the subscription on a monthly basis first to see if it's suitable for you before committing to an annual membership. As an Amazon Prime member, you receive many benefits, including free shipping, video streaming, photo, music, and Kindle books for $119/year in the US and £79 in the UK.

Your Amazon Prime Benefits

Delivery

You get free two-day shipping on eligible items to addresses in the U.S. and free same-day delivery in eligible zip codes.

For Amazon.co.uk, you have unlimited one-day delivery and unlimited same-day (Evening Delivery) on eligible items to eligible locations at no extra cost. In addition, when you select **No-Rush Delivery** on Prime eligible items, you will receive a promotional credit in your Amazon account once the order dispatches.

Videos and Photos

Prime Video gives members unlimited streaming of a huge catalog of Prime movies and TV shows.

You can also subscribe to add-on channels and watch your favorite movies and shows from streaming services like HBO, STARZ, and SHOWTIME, without requiring cable. Amazon Channels costs between $4.99 and $14.99 per month for Prime members.

Photos

Amazon Photos gives you access to unlimited photo storage on your Amazon Drive and 5 GB for video. All Amazon customers get 5 GB of storage. Saving photos to your Amazon Drive enables you to view your photos on a big screen using your Fire TV Stick.

You can also set personal photos as your Fire TV screensaver. See **Customizing Your Screen Saver** in Chapter 10 for how to do this.

Music

Amazon Music is Amazon's music service that allows Amazon Prime customers free access to a selection of music albums. You can create playlists and play music directly from the cloud or download them to your mobile device or PC and play them with the Amazon Music app.

Prime Early Access

Prime Early Access is a feature that gives Amazon Prime members 30-minute early access to Lightning Deals on Amazon. Lightning Deals are items sold at reduced rates and are available until the time the deal expires.

Prime Reading

Prime members have unlimited access to a selection of Kindle books, magazines, comics, and audiobooks. You don't need a Kindle device to access this content. All you need is the Kindle app installed on a compatible device.

Sharing

Amazon Household allows you to share Amazon content with your family members using Family Library. Amazon Household members can share their delivery benefits, Prime Video streaming, and Prime Reading with one adult member of their household.

Chapter 5: The Home Screen

This chapter will cover the menu items on the home screen of the Fire TV Stick. Please note that Amazon regularly updates the interface of Fire TV. This means menu names and settings may be slightly different after a software update. Often, the changes are cosmetic, and the underlying functions remain the same. The instructions in this book will still guide you to the right areas even if the software has been updated and some menu names have been changed.

Home

The home screen provides a top-level view of content mostly customized for you based on your region and viewing habits. The type of lists you'll see here are regularly changed by Amazon, but generally, you'll find the following lists when you scroll down the screen:

- Your recently watched videos.
- Recently used apps.
- Video recommendations based on what you have watched.
- Video recommendations based on the apps you've installed.
- Recommended Prime movies and TV shows.
- Movies you can rent and rental deals.
- Prime recommended documentaries.
- Popular games you can play.

For example, if you have the **Netflix** app installed, then you'll get a recommended list of Netflix movies and TV shows. The **Recently Watched** list gives you a list of your recently viewed movies, TV shows, games, photos, or apps. **Recently Used Apps** allows you to quickly access apps you have used recently.

Tip: To rewatch or continue watching a video you've watched recently, on the home screen, scroll down to your **Recently Watched** list, and you'll see the video on the list. To launch an app you've used recently, scroll down to your **Recently Used Apps** list.

To remove an item from your **Recently Watched** list, navigate to the item so that it is highlighted, press the **Menu** button on your remote, and then select **Remove from Recent** from the Options menu.

Find

The Find screen gives you several options to find movies, TV shows, and apps.

- **Search:** The search screen allows you to search for movies, TV shows, games, apps, music, actors, directors, etc. Use the remote to search for content using Alexa or the onscreen keyboard.

  Tip: Searching can be much easier and faster using the Alexa voice button on your remote. We will be covering this in more detail later in the guide.

- **Library:** The library lists all the items on your **Watchlist** and your **Purchases and Rentals**.

- **Films:** This allows you to browse movies using different categories.

- **TV programs:** This allows you to browse TV programs using different categories.

- **Appstore:** This allows you to find and download apps and browse for apps under several categories.

- **Kids and family:** This option allows you to browse for videos suitable for kids and family viewing.

- **Recommended categories:** This option allows you to browse lists of movies and TV shows organized by category, including horror, westerns, science fiction, drama, thrillers, romantic comedies, war films, cartoons, musicals, and stand-up comedy.

Live

Live is a new menu option on Fire TV for movies and TV shows programmed to run on a schedule. If you have installed any apps that offer live programming, you'll see lists of programs and start times here.

You can also view a list of apps in your region that provide live content, including the news.

You can access the TV Guide for live programs on this screen.

Note: You can view many of these "live" programs on-demand using the on-demand version of the streaming app.

Adding Apps to the Home Screen

Here you get a list of all the apps you've installed on your Fire TV stick and a selection of free and paid apps you can download and install like Netflix, Hulu, YouTube, BBC iPlayer, Curzon, etc. The apps you will find in Your Apps & Channels will be specific to your location.

To add an app to the list of apps on the home screen, do the following:

1. On the home screen, select the **Apps** icon ( ). This takes you to **Your Apps & Channels.**

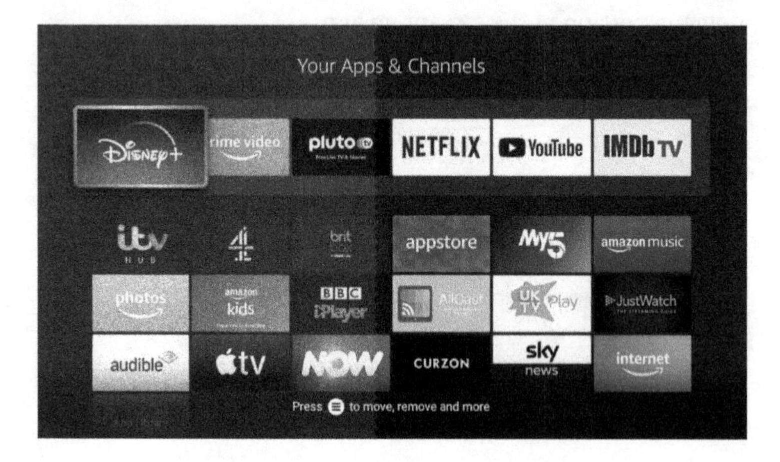

2. In Your Apps & Channels, identify the app you want to move to the home screen from the list of available apps.

   If you haven't installed the app, select **Library** (at the bottom of the screen), and then install the app from the **Appstore.**

3. Use your navigation trackpad to highlight the app and press the **Menu** button to display the **Options** menu (on the lower-right of your TV).

4. On the Options menu, select **Move** or **Move to front** (if you want to move the app to the front of the list above).

5. If you selected the **Move** option in step 4, use the navigation pad on your remote to move the app up to where you want to place it in the list above.

6. Press **Select** on your remote to complete the move.

This will move the app to the list of apps on the home screen.

When you add a new app to the list, Fire TV will move the last app on the shortlist back to Your Apps & Channels.

To remove an app from the home screen, do the following:

1. Go to **Your Apps & Channels.**
2. Select the app in the top row of apps.
3. Press the **Menu** button to display the Options list.
4. Select **Move**, and then use the navigation pad on your remote to move the app down to the list below.
5. Press Select on your remote to complete the move.
6. To go back to the home screen, press **Home** on your remote control.

When you remove an app from the top list, Fire TV will place the topmost app on the list below in the available space.

Settings Overview

Under Settings, you can view and configure many settings in your Fire TV Stick. Note that Amazon is constantly changing the options here, but on most occasions, the changes are usually cosmetic name changes while the underlying function is the same.

Throughout this guide, we'll cover how to make changes in Settings for different features. Below is an overview of the current options in Settings and what they do.

Notifications

Adds notification messages. For example, when you purchase content, you will get a notification stating what content you bought.

Accounts & Profile Settings

Here you can change several settings, including:

- Registering/deregistering your Fire TV Stick.
- Registering a new account.
- Synchronizing your Amazon content.
- Parental controls.
- Managing your profiles.
- Child settings, including adding a PIN.
- Profile sharing.

Network

This option allows you to connect your Fire TV to your Wi-Fi network by choosing from a list of available networks. You also have an option here to save your Wi-Fi password to Amazon.

Display & Audio

Under this setting, you can configure a screensaver and personalize it with your photos. You can also configure several other display and audio settings of your Fire TV Stick.

Applications

On this page, you can configure several options for installed applications. You will see a list of all the apps installed on your Fire TV Stick, and you can manage them in several ways, including deleting cached data if you wish to reset apps.

Equipment Control

Here you can find several options to configure advanced settings for your Fire TV Stick, TV, and any attached devices like speakers. The default setting is Automatic, meaning Fire TV does this for you automatically.

Live TV

Live TV gives you a couple of options that enable you to manage your live TV channels. You can define your favorite channels and set parental controls. You can also sync your live TV sources (apps that you have installed that offer live TV).

Remotes & Bluetooth Devices

Here you can pair, unpair, and update your Amazon fire remote controls. You can also pair game controllers and other Bluetooth devices like speakers, headphones, keyboards, mice, and additional Fire TV remote controls.

Alexa

Under Alexa, you can configure the Alexa app on Fire TV. You also have a **Things to Try** list, which gives you an exhaustive list of commands you can say to Alexa to perform actions on Fire TV for you.

Preferences

Here you can configure several settings for preferences, including:

- Parental Controls
- Privacy Settings
- Data Monitoring
- Notification Settings
- Featured Content
- Location
- Time Zone
- Language

My Fire TV

This option gives you information about your Fire TV Stick, including storage capacity, network details, current version, and when it was last updated. You can also enable some developer options if you wish to sideload apps on your Fire TV stick. You can view the terms of use and other legal notices here. Other actions you can perform here include putting your Fire TV Stick to sleep, restarting Fire TV, and resetting your Fire TV Stick to factory defaults.

Accessibility

This option allows you to set various accessibility options, including subtitles, Alexa subtitles, voice view, text, banner, screen magnifier, high contracts text, audio descriptions.

Help

Under help, you get several help videos which are approximately one-minute videos describing various actions you can perform with Fire TV

Stick, including configuration settings and quick tips for some common issues. You can also send feedback to Amazon using instructions on this page.

Managing Fire TV Profiles

Fire TV Profiles enable you to have a personalized experience for different users on the same Amazon account. With each profile, you can have a personalized watch list, viewing history, recommendations, and watch progress. Profiles also enable you to have personalized apps on the main menu, including some device settings.

An Amazon Kids profile allows you to choose specific apps and content kids can access, set age limits on some content, set limits on screen time, and use a PIN to ensure kids can only use their profile.

Fire TV allows you to have a total of six profiles, of which four can be Amazon Kids profiles.

The content you purchase, rentals, and app subscriptions are shared by all profiles on the same Amazon account.

Note that Fire TV Profiles are not available for the following older Fire TV devices:

- Fire TV (1st Generation)
- Fire TV (2nd Generation)
- Fire TV Stick (1st Generation)

To create a Fire TV profile, do the following:

1. On the home screen, select the Profiles icon.
2. Select **Add Profile.**
3. Select **New Profile.**
4. Next, enter a name and pick an icon for your profile.

To edit a profile, do the following:

1. On the home screen, select the Profiles icon.

   Tip: You can also tell Alexa to take you to the profiles screen by holding down the Voice button and saying, "Go to profiles".

2. Highlight the profile you want to edit.
3. Select **Edit.** This is the pen icon under the profile name.
4. In the Edit Profile screen, you can change the name and icon for the profile.
5. Select **Save** when done. Alternatively, select Cancel if you want to cancel your changes.

You can also manage your Fire TV profiles online with the following weblink:

https://www.amazon.com/manage-your-profiles

To remove a profile, do the following:

1. On the home screen, select the Profiles icon.
2. Highlight the profile you want to remove.
3. Select **Edit.** This is the pen icon under the profile name.
4. In the **Edit Profile** screen, select the **Remove** button.
5. Select **Remove** again to confirm the action.

To create an Amazon Kids profile, do the following:

1. On the home screen, select the Profiles icon.
2. Select **Add Profile.**
3. Select **New profile.**
4. Set "Child Profile?" to Yes.
5. Enter a name, birthdate, and pick an icon for the profile.
6. Select Add.
7. The next step enables you to create a Child PIN. Follow the on-screen instructions to create the PIN.

   The Child PIN ensures kids are kept in their profile and can only see child-friendly content approved by parents.

8. On the next screen, you can add content to the library of the profile. Fire TV pre-selects any child-friendly content you currently own.
9. Next, if you're not subscribed to Amazon Kids+, Fire TV will present you with an option to subscribe. You can select **Subscribe** or **No Thanks.**

To manage the content in an Amazon Kids profile, do the following:

1. From the home screen, go to **Settings** > **Account & Profile Settings.**
2. Select **Child Settings.**

   Note: This setting may be named **Kid Settings** in your region.
3. On the next screen, select the profile you want to update.
4. Select **Add content.**

Reset Your Child PIN on Fire TV

You can reset your Child PIN if you have forgotten it.

Note: Your Parental Control PIN is different from the child PIN. To change your Parental Control PIN, go to **Settings** > **Preferences** > **Parental Controls.**

To reset your child PIN, do the following:

1. When Fire TV prompts you to enter a Child PIN, try entering the PIN.

2. A reset code will appear below the PIN prompt if the PIN is incorrect.

3. On your computer or handheld device, use an internet browser to go to the Amazon Code page:

   https://www.amazon.com/code

If you're not signed in to your Amazon account, you'll need to sign in with your login and password first.

4. In the textbox under Register Your Device, enter the reset code displayed on Fire TV and click **Continue**.

On Fire TV, follow the on-screen instructions to reset your Child PIN.

Chapter 6: Searching for Content

In this chapter, we will cover using the Alexa voice remote control to search for content on your Fire TV Stick. We will also cover the manual method of performing a search using the on-screen keyboard. Alexa should be your first choice for searching for content as it is so much easier, faster, and flexible using voice search on your Fire TV. The manual method using the on-screen keyboard is a good backup in situations where you can't use Alexa.

Using Alexa to Search for Content

The **Voice** button on your remote control enables you to use your voice to interact with Alexa, which is now built into the Fire TV Stick. Alexa can perform searches for content on your Fire TV or return specific information you have asked for, like the weather, for example.

Alexa Voice Remote Overview

To activate the voice search, press and hold the **Voice** button on your remote control and say what you want Alexa to do. You can search for movies and TV shows, ask about the weather, check traffic, launch apps, get information from Wikipedia, etc. Alexa will answer back directly through your Fire TV by displaying the information on the screen, like a list of movies, or through audio, if that would be sufficient.

For example, if I hold down the Voice button and say:

"Wikipedia Apollo 11".

Alexa will show the Wikipedia information from Apollo 11 on the screen as well as narrate it to me. At some point, Alexa will ask if I want more of the information narrated.

When you're done with the information presented by Alexa, press the **Back** or **Home** button on your remote control to return to the previous screen.

Manage Alexa Features

You can manage and configure your customized Alexa settings with a website or a mobile app.

Website

US: https://alexa.amazon.com

UK: https://alexa.amazon.co.uk

App

The app is called **Amazon Alexa**, and you can download it from the iTunes store for Apple devices or the Play Store for Android devices. You can configure the same information you can set with the website above.

A cool feature of the website or app is that it keeps some history of the information you searched for using Alexa in the past. Hence you can access the exact phrase you used in the past to find the same information or content at a later date.

Things to Try

If you want a quick overview of what Alexa can do on the Fire TV Stick, just press the Voice button, and say:

Tell me what I can do.

Alexa tips are also available in the Fire TV under:

Setting > Alexa > Things to Try.

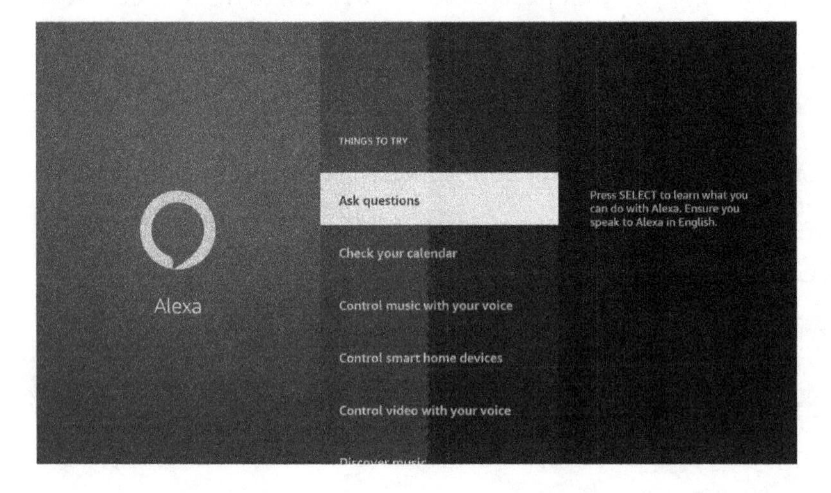

Searching for Movies and TV Shows

To search for movies, TV shows, apps, games, and more, say:

Search for [title / actor / genre / phrase].

Alternatively, you can leave out the "Search for" and just tell Alexa what you want her to search for.

For example:

Search for Daniel Craig.

Or just

Daniel Craig.

Both phrases will return the same results, so you might as well leave out the "search for".

Other examples:

Search phrase	Result
Star Wars	Returns a listing of all Star Wars movies available for viewing on the Fire TV.
Sci-fi movies	A listing of all sci-fi movies.
Historical TV shows	A listing of all historical TV shows on Fire TV.
True stories	Returns a listing of movies and TV shows that are based on true stories.

Note that what you can find with Search will be largely based on the metadata entries for the items. For example, if you search for "historical tv shows", you'll get a list of videos with *Historical* as part of their genre list.

When you search for a phrase like "true stories", the search results will only include movies or TV shows where "true story" is mentioned somewhere in the product description. So, there might be many other videos on the Fire TV that will not be found with that search term if they haven't been appropriately tagged.

Tip: If you are searching for content on a topic or genre, you can try different search terms related to the type of content you're looking for. Each search term may return slightly different results. For instance, if you're looking for **true stories**, you can use the following phrases to generate different results and then add the relevant items to your Watchlist.

True story
True stories
Real story
Real stories
Real life events

How to Find Prime Only content

To find content that is included with your Prime subscription, add the phrase "Prime Only" to your search term or phrase.

Example:

Search phrase	Result
Historical TV shows Prime only	This will return all historical TV shows that are part of Prime.
Sci-Fi TV shows Prime only	This will return all Sci-Fi TV shows that are part of Prime.

Launching Apps with Alexa

You can use Alexa to launch your favorite applications using the **Launch** or **Open** commands.

Open [app name].

Or

Launch [App name]

For Example:
- *Launch Netflix*
- *Open BBC iPlayer*

Search for and Play Music

If you are subscribed to Prime or if you have a standalone subscription to Amazon Music (Amazon Music Unlimited), then you can use Alexa to play music on your Fire TV Stick.

To play music, here are a few examples of what you can say:

Say this	Result
Play music by [artist].	Plays music in your Amazon Music library by the artist.
Play [artist name, album name, song name, playlist] on Amazon Music.	Plays music in your Amazon Music library by artist, song, album, or playlist.
I like this song [Song name]	Adds the song to your library.

Audible & eBooks

You can install the **Audible for Fire TV** app on your Fire TV Stick and listen to Audiobooks.

Once Audible for Fire TV has been installed, to listen to a book, simply say the following to Alexa.

Play the book [title]

Or

Play [title] from Audible

Or

Read [title]

Alexa will launch Audible and play the title you've requested.

Flash Briefing

You can get fast news updates with Flash Briefing.

To get a quick news update, you can say this:

What's my Flash Briefing?

Or

What's going on today?

Or

What's new?

You'll hear a pre-recorded news update that's local to your area from a popular news broadcaster, including the latest headlines and local weather.

For example, if you're in London, you'll get a flash briefing from BBC News that's about 5 minutes long. If you're in the US, you'll get a briefing from your local popular news network.

Weather & Traffic

To check the weather and traffic information with Alexa, you can say the following:

What's the weather like?

Or

Will it [rain, snow] tomorrow?

Or

What's my commute?

What's traffic like right now?

Alexa will tell you information that is local to your area as your Fire TV Stick knows where you are with its built-in Global Positioning System (GPS).

Tip: During the initial setup of your Fire TV Stick, your location would have been set automatically using the home address you've used on your Amazon account. However, if for some reason it wasn't set, you can go to the Fire TV Settings menu (**Settings** > **Preferences** > **Location**) and set it there so that you get customized news, weather, and traffic information.

Q&A and Jokes

You can use Alexa for general information and light jokes.

For example:

What time is it in New York [or any city]?

What's the definition of [word]?

Tell me a joke.

Alexa will tell you a joke.

Alexa may not have responses for everything, but you can help improve Alexa by leaving feedback with Amazon if there is something Alexa could not answer. You can leave feedback under **Help & Feedback** in the web app or mobile app.

Website apps:

US: https://alexa.amazon.com

UK: https://alexa.amazon.co.uk

Wikipedia on Alexa

You can use Alexa to search Wikipedia for information, and Alexa will narrate it for you.

You can say:

Wikipedia [thing].

Alexa will find the requested information from Wikipedia.

Sports Scores

What was the score of the [sports team] game?

For example, if you live in the UK, you can ask:

What was the score of the Arsenal game?

Alexa will give you the result of the last game Arsenal (a soccer team in England) played and the date of their next game if they have one coming up soon.

You can ask:

When is the next [sports team] game?

Alexa will give you the date of the next game for the sports team.

Local Search

Alexa can find local information for you like local businesses, shops, restaurants, etc.

What [businesses/restaurants] are nearby?

Find the hours for [business/restaurant].

Find information about nearby restaurants, shops, and other businesses.

Tip: Alexa uses your postcode or ZIP code in your Fire TV settings and information from other sources like Yelp to find places nearby. It defaults to the postcode or ZIP code on your Amazon account. However, if you want to change this information for even more accurate results, you can update your postcode/ZIP code in your Fire TV settings (**Settings** > **Preferences** > **Location**).

Smart Home Cameras

You can connect a compatible smart home camera to Alexa and then use your voice to access the live camera feed on your Fire TV Stick (2nd Generation).

You can say:

Show [camera name]

To show the live camera feed.

Or

Hide [camera name]

To hide the live camera feed.

Note: How to connect a smart home camera to Alexa is outside the scope of this book. However, for more information on this topic, you can go to the Amazon help page on how to use a smart home camera with Alexa.

US:
https://www.amazon.com/gp/help/customer/display.html?nodeId=202168000

UK:
https://www.amazon.co.uk/gp/help/customer/display.html?nodeId=202168000

Searching with the On-Screen Keyboard

To search for content using the on-screen keyboard, from the home screen of Fire TV, select **Find** and then **Search.**

On this page, you can use the on-screen keyboard to enter your search term. Use the navigation trackpad to move left, right, up, and down on the keyboard. Use the two last characters on the on-screen keyboard (**x**) for backspace and (_) for space. You can also use the **Fast Forward** button for space and the **Rewind** button for backspace.

Searching by Text

The availability of autosuggestions, which display as you type, make searching with the on-screen keyboard easier. You usually do not need to enter full search terms. You just need to enter the first few letters using the on-screen keyboard to get a related list of auto-suggested terms from which you can choose.

For example, if you want to search for "Action movies", you just need to enter "act", and you get auto-suggestions like in the list below:

Act
Love Actually
Action Movies
Action Movies On Prime
Action
Paranormal Activity

You can scroll down this list and select one. You can access more options by scrolling down past the last item on the screen.

You can also search for actor name or director name. For example, if you want to find all movies by "Naomi Watts", you just need to enter the first few letters of her name like "Nao", and a list will be displayed that has "Naomi Watts" as one of the return values.

Nao
Naomi Watts
Naomi Rapace

Naomi
Naomi Watts Movies

When you scroll down and select her name, you will get a list of all movies and TV shows crediting Naomi Watts.

When you find the video you want to watch, you can select **Watch Now** or **Add to Watchlist** to add it to your Watchlist.

To see more viewing options, select the actual artwork of the video to go to the details page. There you will see other options like **Rent**, **Buy**, **Watch Trailer,** and **More Ways to Watch.**

Chapter 7: Watching Movies and TV shows

The Fire TV interface is easy to navigate. Use the navigation trackpad on your remote control to move up, down, right, or left through the items on the screen.

Accessing Prime Content

There are several ways you can find and watch Prime videos.

Home Screen

You can browse Prime Video content from the lists of recommendations on the home screen. Use the navigation pad on your remote to scroll up or down the home screen to see different lists and content categories. The recommendations are mostly tailored to your region, the apps/channels you've installed on Fire TV, and your viewing habits.

If you are a Prime member, look for the "prime" prefix in the list names. Any list with the prime prefix contains movies you can watch at no extra cost.

You'll see **Watch Now with Prime** on the details screen of videos included in Prime.

Using Search

You can access Prime content by using the **Search** function on the **Find** screen. For example, if you are looking for a movie that is part of Prime, you would select **Find** > **Films** and look for the lists with the "prime" prefix.

To look for Prime TV shows, select **Find** > **TV Programs** and scroll down to the lists with the "prime" prefix.

Prime Video App

You can also use the Prime Video app to view Prime content.

Note: If the Prime Video app is not available on your home screen, you can go to Your Apps & Channels and move the app to the home screen. To move the Prime Video app to the home screen, follow the steps detailed in Chapter 5 under **Your Apps & Channels.**

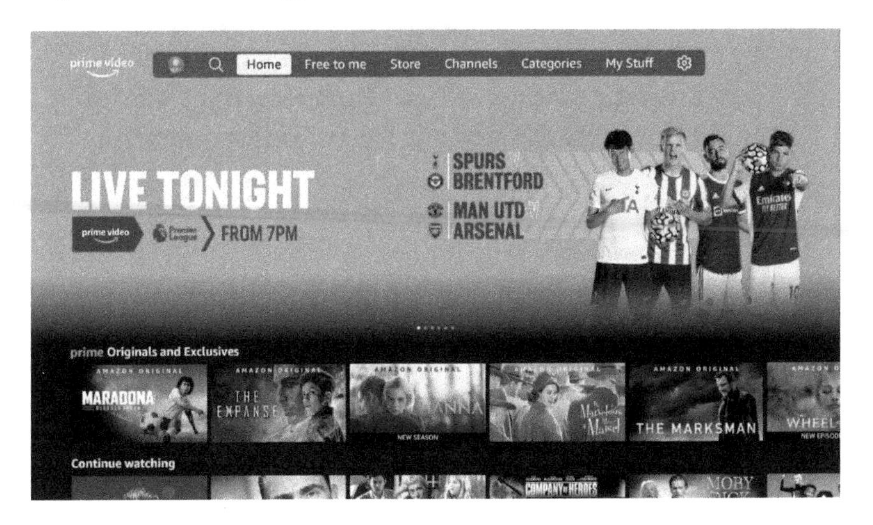

The Prime Video app presents you with a home screen tailored for Prime content.

The main menu of the Prime Video app has the following options:

- **Home:** Allows you to access Prime originals and exclusives and other lists, including your viewing history, so that you can continue watching a video.

- **Free to me:** This option allows you to browse Prime content that you can watch at no additional cost.

- **Store:** You can buy or rent movies here that are not included in Prime.

- **Channels:** Here, you can find a list of several channels you can subscribe to as add-on channels to your Prime subscription.

- **Categories:** This allows you to browse content using different categories. For example, if you want to browse top-rated videos, select **Top rated** on this screen.

- **My Stuff:** This screen has your **Watchlist**, which lists videos you've recently watched and your **Purchases.** If you want to quickly access the TV shows and movies you have rented or purchased, select My Stuff.

Adding items to Your Watchlist

When you come across an item, you would like to save to your **Watchlist**, press the **Menu** button and select **Add to Watchlist** from the pop-up menu on the lower-right of the screen. This enables you to quickly add videos to your Watchlist that you want to watch later without leaving the list you're viewing.

Renting or Buying Movies and TV Shows

If you are not a Prime member or you want to buy or rent videos, then on the home screen, look for the lists prefixed by "Rent or Buy". These lists have titles that you have to buy or rent to watch.

You can also buy or rent videos, including recent releases, from the **Store** menu option in the Prime Video app.

When you buy a movie or TV show, it is added to your video library with lifetime access. When you rent a video, you'll have 30 days to begin watching the video. Once you've started watching it, you have 48 hours to finish watching it before the rental expires.

Your Video Library

After buying or renting a video, Fire TV will add it to your **Library**, which you can access from the **Find** screen.

To watch a video you rented or purchased, from the home screen, navigate to **Find** > **Library**. Scroll down to **Purchases and Rentals** to locate the video you want to watch and select it. On the details screen, select **Watch Now** to begin streaming it or **Resume** (if you previously started watching it).

You can also access purchased and rented videos under **My Stuff** in the Prime Video app.

AutoPlay for TV Shows

AutoPlay is enabled by default in your Prime video **Account & Settings** on the Amazon website. When watching a TV show, as soon as an episode ends, Fire TV will begin the next episode or the trailer of another recommended video.

To turn off AutoPlay, navigate to your **Account & Settings** web page using a web browser on a PC or handheld device. The weblink will vary depending on your country. The links for the US and UK are shown below.

US: https://www.amazon.com/gp/video/settings

UK: https://www.amazon.co.uk/gp/video/settings

On the **Player** tab, set **Auto Play** to **Off**. If AutoPlay has been turned off and you want to turn it on, then select **On.**

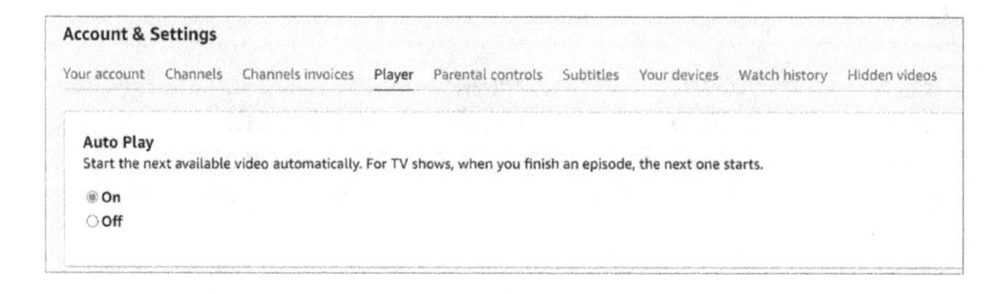

Turn Subtitles On/Off

You can turn on subtitles (or closed captions) on your Fire TV to show the audio for movies and TV shows as text on the screen. Movies and TV shows with subtitles have the "CC" symbol on their details screen. Most videos support English subtitles, but other languages can sometimes be supported too. Netflix videos, in particular, supports several other languages.

To turn on subtitles:

1. Start playing the video.
2. Press the **Menu** button on your remote control to display the Options menu.
3. Select **Subtitles**.
4. If subtitles are available, you'll see an option for **English [CC]**.
5. Select **English [CC]**.
6. On the next screen, you'll see **Off** and **English [CC]** as options. Select **English [CC]** again.

   If subtitles are available for the video, you will get a sample of how the subtitles will look on screen. You can use this screen to change the **Language** (if there is more than one language option).

   On this screen, you can also change the text **Size** and **Style** of the subtitles. For example, you can change the style to yellow text on a black background if the default white text on a transparent background is not suitable for a particular video.

7. Press **Play** on your remote control to resume playback with subtitles from now on.

Note: Your settings here will apply to all videos with subtitles you watch going forward.

To turn subtitles off:

1. Press the **Menu** button while the video is playing.
2. Select **Subtitles** from the Options menu.
3. Select **English [CC]**.
4. On the next screen, you'll see options for **Off** and **English [CC]**. Select **Off**.
5. Press the **Play** button on your remote to resume playback with subtitles now off.

Manage Your Subtitle Settings Online

You can manage your Amazon Video subtitle settings in your Prime video **Account & Settings** online under the **Subtitle** tab.

US: https://www.amazon.com/gp/video/settings/subtitles

UK: https://www.amazon.co.uk/gp/video/settings/subtitles

Netflix Captions

To set how subtitles appear when you watch Netflix, visit **Subtitle Appearance** at **www.netflix.com/youraccount**. You can also update how subtitles appear within other video applications within those applications, for example, BBC iPlayer.

Use X-Ray to Learn More About Videos

X-Ray is a feature available on Fire TV and Kindle Fire devices that enables you to learn more about the video you're watching. It helps to enrich your experience by embedding Internet Movie Database (IMDb) information into the viewing experience. X-Ray for movies and TV shows enables you to learn information about the actors in particular scenes, the soundtrack, and other trivia.

You can learn more about X-Ray from the IMDb website:

https://www.amazon.com/adlp/xray

Note that X-Ray is not available for all content. To find out if X-Ray information is available for a particular video while it is playing, press the top of the navigation trackpad on your remote (**Up**). If it is available, you'll see **View All X-Ray** next to **Options** at the bottom-right of the screen.

Display X-Ray Quick View

The quick X-Ray screen allows you to see information on-screen, such as the actors in the current scene.

To display X-Ray Quick View:

1. While the movie is playing, press **Up** on your remote control.

   This will bring up the **X-Ray Quick View** screen while the movie continues to play in the background. You'll see information at the bottom of the screen relating to the current scene.

2. To hide X-Ray Quick View, press **Down** on the navigation trackpad on your remote.

Note: You can also press **Play/Pause** on the remote to display the **X-Ray Quick View**. Press **Play/Pause** again to resume playback.

Display the Full X-Ray Screen

To see in-depth information about cast, music, and trivia, do the following:

1. While a video is in playback, press **Up** twice on your remote to display the X-Ray screen. The movie will be paused in the background when you do this.

2. Use the navigation trackpad on your remote to move through the menu and select the menu items you want to view.
3. Press **Play/Pause** on your remote control to close the X-Ray screen and resume playback.

Skip to Specific Scenes Using X-Ray Information

You can use X-Ray to find and skip to specific scenes in a video. X-Ray information also includes a breakdown of scenes and a short description of what the scene is about.

To find and skip to a particular scene in a video, follow the steps below:

1. While a video is in playback, press Up twice on your remote control to display the X-Ray screen.

2. On the X-Ray screen, select **Scenes** from the top menu.
   A series of scenes will be displayed at the bottom of the screen. For some videos, each item has a short description of what that scene is about.

3. To select a particular scene, use your navigation trackpad to scroll to it and press Select on your remote control. Video playback will be resumed from that point.

Watching Videos in 4K

To watch videos labeled as UHD (Ultra High Definition) in 4K, you need a Fire Stick 4K, a 4K TV, and an Internet connection of at least 25 Mbps. If you are using a wireless connection, the Wi-Fi frequency you use will also play an important role.

Use the 5 GHz Wi-Fi Frequency Band

Wi-Fi travels on frequencies called radio bands, and modern-day routers capable of streaming 4K videos are usually dual-band or tri-band. Dual-band routers have 2.4 GHz and 5 GHz frequencies, while tri-band routers add a third band which is another 5 GHz frequency.

To watch UHD videos on your Fire TV Stick 4K, ensure your Wi-Fi connection is using the 5 GHz frequency (only worry about this if you are using a wireless connection). If your Wi-Fi router has a 5 GHz frequency option, you'll see it as an option in the list of available networks when you're setting up the Wi-Fi connection for your Fire TV Stick 4K. Select it for your Wi-Fi connection.

A video that is available in 4K will have the **UHD** (Ultra High Definition) marking on its details page. See the image below.

To check if 4K playback is possible with your setup (i.e., your Fire TV Stick, TV, and Wi-Fi connection), start playing the video, then press the **Menu** button on your remote control.

This will display the **Options** menu on the lower right of the screen. You should see **4K UHD playback details** among the menu items if 4K playback is available for that video.

To check whether the video is playing in 4K, press the bottom of the circular trackpad to display some video details at the bottom of the screen. The playback mode should be displayed as **Ultra HD HDR.**

Note: A UHD video may initially start in HD mode before eventually switching to Ultra HD HDR mode. This process may take a few seconds to a few minutes, depending on the strength of your internet connection signal and if you have previously been watching the video in 4K.

Set Viewing Restrictions

You can set viewing restrictions on your Fire TV so that a PIN is required to watch videos above a certain content rating. You can do this by enabling Parental Controls on your Fire TV.

To turn on Parental Controls:

1. From the home screen, select **Settings** > **Preferences** > **Parental Controls**.
2. Select **Parental Controls**.

   Note: If you already have Parental Controls turned on, you'll be prompted to enter your PIN. Otherwise, you'll see an option to turn Parental Controls ON.

3. Follow the on-screen instructions to enter your **Prime Video PIN** (or create a new one if you don't have one already).

   Entering your PIN turns on Parental Controls, and Fire TV takes you back to the Parental Controls menu, where you'll now have more menu options.

4. Select **Viewing Restrictions** in the Parental Controls menu.

   Under Viewing Restrictions, you'll see the rating categories you'd like to restrict for your Fire TV.

   There are currently five video categories you can restrict with your PIN.

   - **General** – Suitable for all ages. Ratings: U, ALL
   - **Family** – Suitable for ages 6 and above. PG, 7+
   - **Teen** – Suitable for teenagers. Ratings: 12, 13+
   - **Young Adults** – Suitable for Young Adults. Ratings: 15, 16+

- **Mature** – Suitable for Adults.

5. Scroll down to the option you want to change, and press **Select** on your remote to lock (or unlock) the option. A **lock sign** means a category is restricted and requires the PIN to watch. A **checkmark** means the category is not restricted.

Note: These restrictions are only for Prime videos. If you're subscribed to other networks, like Netflix, you'll need to use the settings/preferences within that app to set any viewing restrictions you want to set for videos from that app.

After you set viewing restrictions (a **lock** symbol) and the message *"viewing restrictions apply"* will be displayed next to titles with ratings in your restricted categories. When you try to watch restricted titles, you'll be prompted to enter your PIN.

Important: Ensure you remember your PIN once set because you will need it each time you want to set Parental Controls. Even if you turn off Parental Controls, your Amazon account still stores your PIN, and you'll need it if you want to reactivate Parental Controls. Entering a different PIN will not work if the previous PIN is still on your Amazon account.

If you cannot remember your PIN, visit your Prime video **Account & Settings** to reset your PIN. Depending on your country of residence, the site will vary, but for US and UK viewers, you can use the links below to access your Prime Video Settings.

US: https://www.amazon.com/gp/video/settings/parental-controls

UK: https://www.amazon.co.uk/gp/video/settings/parental-controls

How to Set Your Prime Video PIN

To place viewing and purchase restrictions on your Fire TV, you'll need to set a PIN for your account. Once the PIN is set, you'll need it to authorize purchases or to bypass viewing restrictions you've placed on certain content.

To create or change your Prime Video PIN:

1. Use a web browser on a PC or handheld device to navigate to your Prime video **Account & Settings** page (see the link below).

2. Click the **Parental Controls** tab.

3. Enter a 4-digit number to set up your PIN.

   If you have a PIN that you want to change, click **Change** to change your current PIN.

4. Click **Save** when done.

US: https://www.amazon.com/gp/video/settings

UK: https://www.amazon.co.uk/gp/video/settings

Set Purchasing Restrictions

Similar to viewing restrictions, if there are several people in your household, you may want to put purchasing restrictions on your Fire TV, which is registered to your Amazon account. You can do this by enabling Parental Controls for purchases so that a PIN is required.

To turn on Parental Controls:

1. From the home screen on Fire TV, select **Settings** > **Preferences** > **Parental Controls**.

2. Select **Parental Controls** again and follow the onscreen instructions to enter your Prime Video PIN (or create a new one if you do not have one already).

   Note: If Parental Controls is already ON, you'll be prompted to enter your PIN.

3. Ensure that **PIN-Protect Purchases** is set to ON.

Once the **PIN-Protect Purchases** has been set to ON, whenever you attempt to buy or rent a video, a game or paid app, you'll be prompted to enter your PIN.

Tip: If you want to minimize impulsive purchasing, you could set this option even if you're the only one using your Fire TV.

You can also use the PIN to restrict launching apps and viewing photos by setting:

- **PIN-Protect App Launches** to ON.
- **PIN-Protected Amazon Photos App** to ON.

Creating a Watch Party

You can have a watch party where several people in different locations can watch videos on Fire TV at the same time using the Watch Party feature. The watch party feature is available from the Prime Video app on Fire TV or the Prime Video app on your mobile device or computer.

To start a Watch Party, do the following:

1. On Fire TV, open the Prime Video app and select the video you want to watch.

2. On the details page of the video, select **Watch Party.**

3. You will get two options on the next screen - **Create Watch Party** or **Join Watch Party.**

4. Select **Create Watch Party.**

   Fire TV gives you two ways you can create a chat room. You can use a unique web link or a QR code.

5. Enter the provided web link in a web browser on your mobile device or computer. Alternatively, you can scan the provided QR code. You can use your camera app or a QR code reader app on your mobile device.

   This will open a window in the Prime Video app that enables you to join the chat room.

6. In the Prime Video app on your mobile device or computer, enter the name you want to use in the chat room (this can be an alias or nickname), and then select **Join Watch Party**.

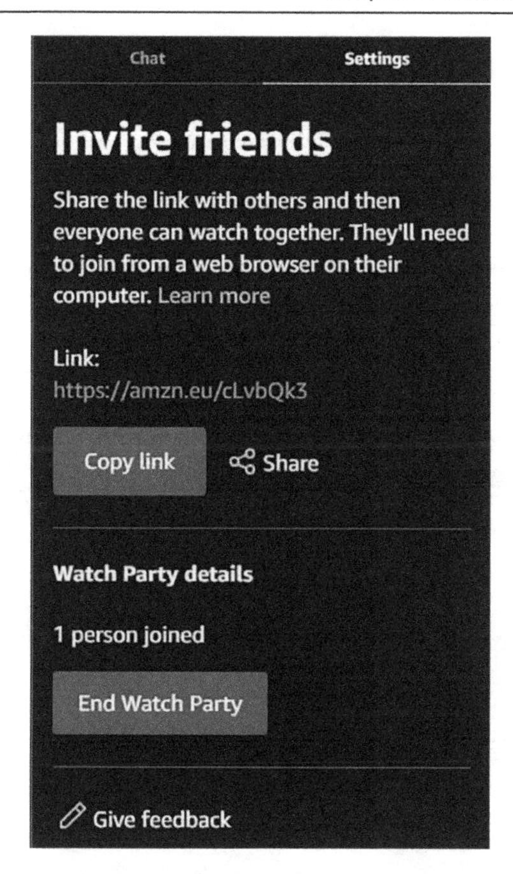

Amazon Video opens the chat room where everyone in the watch party can chat and discuss the video you are watching.

7. On your Fire TV stick, select **Start Watching** to begin watching the video.

8. To invite other people to the watch party, send them the web link using your mobile device or computer.

Fire TV allows you to invite up to 100 people to a watch party. Anyone with a valid link can join the watch party, but they should be located in your country and have access to the video you're watching through their Prime membership.

Monitoring Data Usage

Amazon Fire TV provides a tool that enables you to monitor how much Internet data your device is using and how that data is split between the apps.

The data monitoring tool enables you to keep track of your data consumption if that is necessary. This is especially useful if you have a monthly limit on your Internet data usage, or perhaps you just want to know which apps are using the most data.

Note: The data monitoring tool can only provide information on data usage. You can't use the feature to restrict how much data your Fire TV uses.

To turn on data monitoring, do the following:

1. Navigate to **Settings > Preferences > Data Monitoring**.

2. Then, press Select on your remote control to turn data monitoring **ON** or **OFF**.

Data Monitoring Preferences

Once you've turned Data Monitoring on, you can manage the following options:

- **Set Video Quality**

  This option allows you to adjust the quality of your video streams. You can choose between three options – **Good**, **Better**, and **Best**. The higher the quality, the more data Fire TV will use.

- **Set Data Alert**

  This option enables you to monitor the amount of data you are using each month and to display notifications when you have reached your limit. You can specify the data limit in gigabytes, and the monthly period you are monitoring (for example, 30 GB, from the 1st of each month).

  On-screen notifications will automatically be displayed once your Fire TV has reached your specified data limit. You will get an alert when you reach 90% and another when you have reached 100% of your data limit.

- **Monthly Top Data Usage**

  This option allows you to view data usage for your Fire TV apps. You see a listing of each app and how much data it has used this month. This enables you to keep track of which apps are using the most data, so you know what applications to monitor if you want to keep your data usage down. You can also view details of how much data each app used in the previous month.

Tip: You only really need to worry about monitoring data usage on your Fire TV if you have a monthly data usage limit on your Internet service. If you have unlimited data usage, it is unnecessary to turn on Data Monitoring unless you have other reasons to do so.

Tip: To get the best experience from your Fire TV, always leave the **Set Video Quality** option at the default setting of **Best**. Only change this setting if you have limits on how much data you can use each month.

Mirroring Your Mobile Device

You can mirror the display of most phones or tablets that are Miracast capable on your TV using your Fire TV Stick. Miracast is a peer-to-peer technology that allows wireless connections using Wi-Fi Direct. It allows the sending of up to 1080p HD video and 5.1 surround sound between devices.

Devices that are Miracast capable include:

1. Fire phone (now discontinued).
2. Fire HD/HDX tablets.
3. Devices running Android 4.2 (Jelly Bean) or higher.

Before you begin, make sure your Miracast capable device and Fire TV Stick are within 30 feet of each other, meaning you should not be too far from your TV. Also, if you are using a device that is not registered to the same Amazon account as the Fire TV, ensure your device is connected to the same Wi-Fi network as your Fire TV.

To start Display Mirroring:

1. Press and hold the **Home** button on your Fire TV remote.
2. Select **Mirroring** from the available options.
3. Then connect your compatible device.

How To Connect Your Devices

Compatible Fire Tablet:

1. Swipe down from the top of the screen to open Quick Settings, and then tap Settings.
2. Tap Display & Sounds, and then tap Display Mirroring.

3. Select your Fire TV Stick. It may take up to 20 seconds for your Fire Tablet screen to appear on your TV screen.
4. To stop mirroring your Fire Tablet, tap Stop Mirroring.

Android device running Android OS 4.2 or higher:

1. On your Fire TV, select **Settings** > **Display and Audio** > **Enable Display Mirroring**.

   Alternatively, you can press and hold down the **Home** button and select **Mirroring** from the Quick Access menu.
2. On your Android device, connect to your Fire TV Stick. The method you use may vary depending on your Android device and the version of your Android OS. Consult the help guide for your device for how to do this if necessary.
3. To stop Display Mirroring, press any button on the remote.

Fixing Audio Sync Issues

Occasionally, you may experience audio sync issues when using your Fire TV Stick, where the audio no longer synchronizes with the video. Audio sync issues can happen when running third-party apps like YouTube, Netflix, and other video streaming apps. Just turning the device off and on does not fix the problem, so this can be frustrating. Thankfully, there is a fix, and we'll go through the troubleshooting process in this section.

Follow the steps below to resolve audio sync issues with your Fire TV Stick:

Step 1: Configure Display & Audio

1. From the home screen, navigate to **Settings** > **Display & Audio.**
2. On the **Display & Audio** menu, scroll down and select **Audio.**
3. On the Audio screen, select **Advanced Audio** and ensure **Volume Leveler** and **Dialogue Enhancer** are both set to **OFF.**
4. On the Audio screen, select **AV Sync Tuning.** Move the slider left or right with your navigation trackpad on your remote, and press select until the flashing circle is in sync with the sound.
5. Select **Apply** if you're happy with the sync.
6. Use the Back button to return to the **Settings** menu.

Step 2: Delete the cache for all Managed Applications

1. On the **Settings** menu, select **Applications** > **Managed Applications.**
2. For each app in the Managed Applications list, select the app and then select **Clear cache** on the next menu.
3. Repeat this for all the apps on the list (or those that you use). Just make sure you don't click on *Clear data* for any of them.

4. When you're done, use the Back button on your remote control to return to the **Settings** menu.

Step 3: Restart Fire TV

1. From the **Settings** menu, select **My Fire TV**.
2. Scroll down and select **Restart** from the menu, and select **Restart** again to confirm the action.

The restarting action will reboot the Fire TV Stick. This is an important step in the process. Otherwise, the audio sync issue may not be fixed.

When Fire TV restarts, the sync issue should have been resolved.

Chapter 8: Pairing Bluetooth Devices with Fire TV

You can connect other devices to your Fire TV Stick, for example, game controllers, additional Fire TV remote controls, Bluetooth headphones, and Bluetooth speakers. In this chapter, we will cover how to pair Bluetooth sound devices and additional remotes to your Fire TV.

Pairing Fire TV with a Bluetooth Speaker

Being able to connect your Fire TV directly to a sound device is useful if you're using a PC monitor (without built-in speakers) with your Fire TV Stick, or perhaps you may want to use headphones. Also, you may prefer the sound quality of your Bluetooth speakers over that of your TV speakers.

Note that when you pair the Fire TV Stick with a Bluetooth sound device, it no longer sends sound to your built-in TV speakers. The sound is diverted to the paired sound device as Fire TV can only transmit sound to one device at a time.

To connect a Bluetooth sound device to your Fire TV Stick, do the following:

1. From the home screen, select **Settings** > **Remotes & Bluetooth Devices** > **Other Bluetooth Devices.**

2. Select **Add Bluetooth Devices**.

3. Next, refer to the user manual of your Bluetooth device to put it into **pairing** mode (there would usually be a flashing light on your Bluetooth device to indicate it is in pairing mode).

   When Fire TV detects your Bluetooth sound device, the name will appear under **Add Bluetooth Devices**.

4. Under **Add Bluetooth Devices**, choose the device and press **Select** on your remote to pair it with Fire TV.

Notes:

- If you're having trouble pairing your device, make sure you unpair it from other devices first before attempting to pair it with Fire TV.

- At the time of this writing, the volume controls on the Alexa Remote Control do not control the volume on a paired Bluetooth sound device. You must use the volume controls on the Bluetooth sound device to adjust the volume. Hopefully, this functionality will be added to Fire TV in the future.

Unpairing a Bluetooth Device

Once you've paired your sound device to your Fire TV, any time you turn it on, it will automatically connect to your Fire TV. For some Bluetooth sound devices, you'll need to unpair them from one device before you can use them with another device. For instance, if you have a Bluetooth speaker paired with Fire TV and you want to use it with your smartphone, you may need to unpair it from Fire TV first to put it back into pairing mode and available to be paired with your smartphone.

Note that when you unpair a Bluetooth device from Fire TV, it removes it completely. You'll need to go through the process of pairing it again if you want to use it with your Fire TV again.

Tip: If you do **not** want to unpair your sound device from Fire TV, but you want to make it available for pairing with another Bluetooth device, simply **turn off the Fire TV Stick at the mains** before turning on the Bluetooth device. This will put it into pairing mode (as it can't find the Fire TV) and is available for pairing with another Bluetooth device.

To unpair a Bluetooth device, do the following:

1. Navigate to **Setting** > **Remotes & Bluetooth Devices** > **Other Bluetooth Devices**.
2. Choose the Bluetooth device you want to unpair from the list.
3. Press the **Menu** button on your remote to unpair it.
4. Press the **Select** button to confirm the action.

That's it. Your Bluetooth device will now be free and available to be paired with another Bluetooth device.

Chapter 9: Installing Apps

In this chapter, we will cover how to download and install free and paid apps and games on your Fire TV Stick.

To buy and download games and apps from Amazon, you'll need to set up a 1-click payment method on your Amazon account.

To install apps and games, do the following:

1. On the home screen, select **Apps**.
2. Scroll down to the bottom of the **Your Apps & Channels** screen and select **App Library**.
3. Press **Up** on your remote to display the main menu of the Appstore.

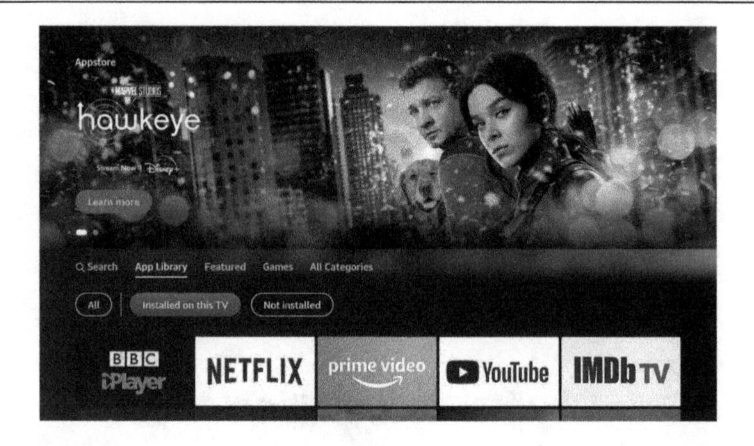

The main menu offers several ways you can find and install apps:

- **Search**: You can use the **Search** feature to find and install apps. You can also use Alexa voice commands.

- **Apps Library**: This allows you to view all the installed apps on Fire TV and the apps previously installed. This option lets you determine which apps you have installed and a quick way to reinstall an app.

- **Featured**: Gives you several lists of recommended apps for different categories.

- **Games**: This allows you to browse for games under several categories such as Action, Adventure, Arcade, Board, Card, and Casino.

- **All Categories**: Use this option to browse apps filtered by category.

4. Once you identify the app or game you want to install, press **Select** on your remote to go to its details screen.

   Some apps and games are free, while others are paid apps. If an app is free, you'll see **Free to download** next to the download button. For paid apps, you'll see the prize next to a shopping cart.

5. To view **customer reviews** for the app or game, scroll down the screen to the **Customer Reviews** section. Customer reviews may help you decide if the app meets your requirement, especially a paid app.

   At the bottom of the details screen, you will also find information on the app's current version, developer, and permissions.

6. Select **Get** or the shopping cart to download the app. If it is a paid app, confirm the purchase on the next screen.

   Important: Purchases are not refundable, so be sure you want an app or game before purchasing it.

Note: For paid apps, you can use your 1-Click payment method, or if you have Amazon Coins available on your account, you can use that.

After the app is downloaded, you'll notice the **Open** option will be in place of **Get.** Select **Open** to launch the app or game.

The apps and games you have installed on your Fire TV Stick can be found in **Your Apps & Games** from the home screen. Click on the **Home** button on your remote control to go to the home screen and select the **Apps** button.

Turning Off In-App Purchases

Some games and apps offer the option to buy in-app items with real money that you can use to unlock additional functionality like new levels in a game or buying subscriptions in an app. You can disable this option on your Fire TV if you do not want accidental purchasing of items.

To turn off In-App Purchases:

1. From the main menu, select **Settings** > **Applications** > **Appstore.**

2. On the next screen, set **In-App Purchases** to **OFF.**

Note: This is more like a measure to prevent accidental purchases. Anyone using the device can go into **Settings** and turn this setting back on. If you want to restrict in-app purchasing with a PIN, you will need to set that in **Parental Controls**. See Chapter 7, under **Set Purchasing Restrictions**, for how to do this.

Uninstall an App or Game

You can uninstall games and apps from your Amazon Fire TV Stick at any time. The games and apps you have purchased will remain stored in the cloud in your account. So, if at any point you wish to reinstall an app or game, you don't have to buy it again. You can simply reinstall it from the cloud.

To uninstall a game or app:

1. From the **Home** menu, select **Settings** > **Applications** > **Manage Installed Applications**.
2. Scroll down to the app you want to uninstall, and press Select on your remote control.
3. Select **Uninstall** from the following menu and then follow the onscreen instructions.

To uninstall an app from your Apps Library, do the following:

1. Use your remote to move to the app so that it is selected.
2. Press the **Menu** button on your remote to display the Options menu on the lower-right of your TV.
3. Select **Uninstall** from the menu and select **Uninstall** again to confirm the action.

Reinstalling an App or Game

The apps and games you have installed are automatically saved to your Amazon cloud account. If you uninstall an app or game, you can reinstall it from **Your Apps & Games**.

To reinstall apps:

1. Press and hold the **Home** button to display the Quick Access menu.
2. On the Quick Access menu, select **Apps**. This will display the **Your Apps & Games** screen.
3. Select an app to reinstall it on your Fire TV Stick.

Chapter 10: Photos and Personal Videos

In this chapter, we will cover how you can view photos and personal videos with your Fire TV Stick. First, we will discuss Amazon Drive as this is where you will be storing your personal photos and videos that you can view on Fire TV.

Amazon Drive Overview

Amazon Drive is a cloud storage service for Amazon customers. All Amazon customers start with 5 GB of free storage for photos, videos, documents, etc. Owners of Amazon handheld devices, like a Fire phone or one of the latest generation Fire tablets, get free unlimited storage for all the photos taken with their device.

Your Prime membership also comes with **Amazon Photos**, which gives you unlimited free storage on Amazon Drive for your photo collection. Thus, with Amazon Photos, the photos you upload to Amazon Drive will not count against your Amazon Drive storage limit.

You can also store videos and other files that are not recognized as photos, but they will count against your Amazon Drive storage limit.

You can upload files to your Amazon Drive directly using your internet browser. Alternatively, you can download and install the Amazon Drive desktop app (for PC or Mac). This enables you to store files in a directory on your hard drive, which is then synchronized with your Amazon Drive online.

To access your Amazon Drive or to download the Amazon Drive app go here:

https://www.amazon.com/clouddrive

Note: If you are not in the US, substitute **Amazon.com** for your home Amazon store.

You can also upload files to your Amazon Drive directly from your mobile device using the Amazon Drive app for handheld devices.

Amazon Photos App

With an Amazon account, you get 5 GB free. If you want additional storage, there are several premium plans from 100 GB to 30 TB to which you can subscribe.

The Amazon Photos app allows you to view your photos and personal videos that are under 20 minutes long with your Fire TV Stick. You can launch the Amazon Photos app from **Your Apps & Games.**

When you open the Amazon Photos app on Fire TV, you will see the following menu:

- **Your Photos** – You can scroll right or left to view thumbnails of all the photos and videos you have uploaded to your Amazon Drive. To view a photo or video, select it, and it will be displayed in full. To return to the thumbnails, press the **Back** button on your remote control.

- **Videos** – The videos you have uploaded to your Amazon drive can be accessed here. Select a video to start playing it. You can use your Fire TV remote control to play, pause, rewind, and fast forward.

- **Albums** – You can create albums using the Amazon Photos website or the Amazon Photos mobile app. You can view your photos organized into albums using your Fire TV Stick.

- **Folders** – Choose a folder to view its contents. You can create and manage your folders using the Amazon Drive website or the mobile app.

Favorites

You can view all the photos or videos added to your favorites list from this list.

To add an item to this list, do the following:

1. Select the photo or video to view it on-screen.
2. Press the Menu button on your remote control. This displays the Options pop-up menu.
3. Select **Add to Favourites** from the Options menu to add it to your Favourites list.

When viewing a collection of photos, for example, a group of photos in a folder, you can press Menu/Options on your remote and choose **Set as Screensaver** from the menu options. This will use the images in the folder as your screensaver from then onwards.

Manage Access to Your Photos and Videos

You can protect your personal videos and images from being viewed by anyone using your Fire TV Stick.

Tip: One way you can restrict access to your photos is to use the **PIN Protect Amazon Photos App** setting in **Settings** > **Preferences** > **Parental Controls** to restrict access to the Amazon Photos app on your Fire TV.

Another way to prevent your photos from being available to your Fire TV Stick is to disable access to your photos in settings.

To manage the visibility of your photos on your Fire TV Stick, do the following:

1. From the main menu, select **Settings** > **Applications** > **Amazon Photos.**
2. Select **Access Amazon Photos.**
3. On the next screen, select **Disable Amazon Photos** (or **Enable Amazon Photos)** and confirm your selection.

When you select **Disable Amazon Photos**, content from Amazon Photos (online) will not be downloaded to your Fire TV. Hence your photos and videos will not be available when you open the Amazon Photos app.

Adding Photos & Personal Videos to Amazon Photos

The easiest way to add photos and personal videos to your Fire TV is to use the web-based Amazon Drive. To access Amazon Drive online using a computer or handheld device, type "Amazon Drive" into your internet browser's search engine and press **Enter.** One of the first results will be a link that will take you to your Amazon Drive for the Amazon store in your country.

The instructions in Amazon Drive for uploading files are self-explanatory and easy to follow.

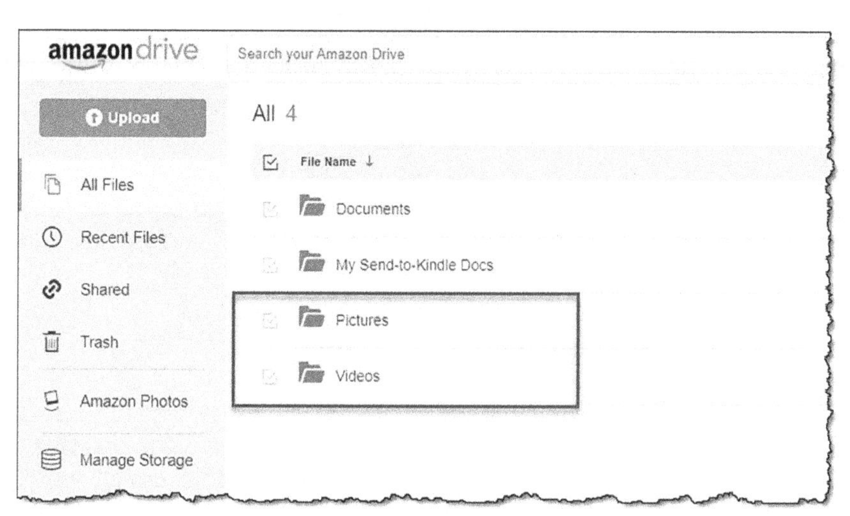

Another way you can upload files is to use the Amazon Photos website.

You can also install the Amazon Photos app on your mobile device from the apps store of your phone. For the desktop version of Amazon Photos, you can download and install the app from the link below.

https://www.amazon.com/Amazon-Photos

Viewing Slideshows

You can generate a slideshow from your photos on your Fire TV Stick. To do so, do the following:

1. From the home screen, select the **App** button to go to **Your Apps & Channels.**

2. In Your Apps & Channels, find and select the Amazon Photos app (named **photos**).

3. Select **View All** to display thumbnails of all your images. Or, if you have created folders, select **Folders** on the menu, and then select the folder you want to view.

4. With the photos displayed on-screen as thumbnails, press **Play/Pause** on your remote control to start the slideshow.

   The slideshow will start and automatically advance through all your photos or personal videos or the items in the selected folder. When it gets to the end, it will restart.

5. You can use left and right on the navigation trackpad on your remote to view the slideshow at your own pace. To pause the slideshow, press **Play/Pause**.

6. To end the slideshow at any point, press the **Back** button.

Tip: While the slideshow is running, to manage slideshow settings, press the **Menu** button on your remote control. This gives you the option to adjust the slide style, slight speed, and toggle shuffle on/off while previewing the slideshow in the background. To exit slideshow settings, press the Menu button again.

Customizing Your Screen Saver

You can display your photos as the screensaver that appears after a few minutes of your Fire TV being idle. So, instead of the default screensaver, you can use your own collection of photos.

To use your photos as your Fire TV screensaver, do the following:

1. In the Amazon Photos app on Fire TV, select View All to display thumbnails of all your images. If you have created folders, select Folders on the menu, and then select the folder you want to view.

2. On your remote, press **Menu/Options** ⊜ and then choose **Set as Screen Saver.**

3. On the next screen, you can configure screensaver settings like **Slide Style**, **Slide Speed**, **Start Time**, **Date and Clock**, **Smart Captions**, **Alexa Hints**, and **Shuffle.** While you do this, you can be previewing the screensaver in the background.

4. Once set, the screensaver will automatically start when your Fire TV has been idle for a few minutes.

Tip: Another way to manage your screensaver settings is to select **Settings** > **Display & Audio** > **Screen Saver** from the home screen. You can adjust various screensaver settings there to your liking.

If you are having trouble with displaying the screensaver, check that the photos from the collection or folder you have set for your screensaver are still available to Amazon Photos. A quick way to check this is to check the **Pictures** folder in your Amazon Drive with a web browser. If the photos are there, but you're still having problems getting the screensaver to work, then restart Fire TV by unplugging the device from the power source and plugging it back in.

Reverting to the Default Screen Saver

To change the screen saver back to the Fire TV default, do the following:

1. From the home screen select **Settings** > **Display & Audio** > **Screen Saver**.

2. Under Screen Saver Settings, select **Current Screen Saver**.

3. Under Current Screen Saver, select **Amazon Collection**.

4. Press the Home button to go back to the Fire TV home screen.

Note: You can select any other option on the list if you want to use a different collection of images on your Fire TV Stick as your screen saver.

Chapter 11: Listening to Music

You can browse and stream music from **Your Music Library** with your Fire TV. These include songs you purchased from Amazon's **Digital Music Store**, imported to Your Music Library, or **Amazon Music** content in Your Music Library. To stream Amazon Music with your Fire TV Stick, it must be authorized to your Amazon account. You can have up to 10 music streaming devices authorized to your account, and each device can only be authorized to one Amazon Music account at a time.

Amazon Music

If you are an Amazon Prime member, you have access to **Amazon Music** as part of your membership, and you can listen to over a million songs, hundreds of playlists, and personalized stations at no additional cost. You can get access to other types of audio content like MP3 books and courses. Not all music on Amazon is included in Amazon Music, but the selection is decent.

To find out more about Amazon Music, go here:

https://www.amazon.com/music/prime

If you are not a Prime member or you want to buy songs that are not available as part of your Prime membership, you can do so from the Digital Music Store. The music you buy will be added automatically to your library.

At the time of this writing, the Digital Music Store is not available within the Amazon Music app for the Fire TV, so to buy new music, you have to use the Amazon website or a hand-held device like a phone or a tablet. However, the music you have purchased and added to your library will be available on your Fire TV for streaming.

Amazon digital music purchases are stored free on your Amazon Drive. You can also import up to 250 songs from your computer into your library for free, including music purchased from iTunes.

How to Play Music on Your Fire TV

Press and hold the Home button on your remote to display the Quick Access screen. Then choose the **Apps** option.

This will open **Your Apps & Channels.** The Amazon Music app is pre-installed, so you can just select it to launch it.

Once you've opened the Amazon Music app, you will see a menu on top of the screen, which you can use to navigate the app and find the music you want to listen to.

Amazon Music menu options:

- **Home** – You can use browse to view music by category, including titles that are available with Amazon Music or Music Unlimited.

- **My Music** – This will display all the music that you've purchased or saved in your music library. The music is organized by category (Artists, Albums, Playlists, Genres, and All Songs).

- **Recents** – This is where you'll find the music you have recently added or recently played.

- **Search** – you can search for music using the title or the genre. When you search, you will get results from **My Music** as well as titles from **Amazon Music** if you are currently subscribed to Amazon Prime.

Tip: If you have the Alexa voice remote control, you can use Alexa to search for music that you want to listen to in Amazon Music. Press the **Voice** button on your remote and say what you're searching for. On the results page, below the search results, you will see a section titled **Search in Apps.** Select **Amazon Music** to get search results related to that app. Once you start playing a song, the Amazon Music app opens, displaying a "Now Playing" screen with a playback progress bar and the lyrics of the song (if they are available).

Listening to Free Music with Stations

Your Fire TV Stick allows you to listen to free music from several music stations covering a wide array of genres.

To see all the available music stations on Fire TV, do the following:

1. Open the Amazon Music app.
2. On the main menu of Amazon Music, select **Search.**
3. On the search screen, select **Stations.**

   This will display a list of stations from which you can stream live music.

4. Scroll through the list, and when you find a station you like, select it. Fire TV will start playing the music on the Amazon Music playback screen.

Note: You can use the Play/Pause button to control playback and the Rewind and Forward buttons to go to the previous song or the next song. There are limits to how many songs you can skip with the free service. If you want to be able to skip several songs on a station, you will need to subscribe to Amazon Music.

Controlling Playback

Use the media buttons on your remote control (**Play/Pause**, **Fast-Forward**, and **Rewind**) to control music playback within the app.

If additional options are available, you will see a prompt on the screen to press the **Menu** button. When you press **Menu**, you see additional options like shuffle, repeat, view album, and more.

Playing Music in the Background

While playing music, you can exit the "Now Playing" screen by pressing **Back** or **Home** to go back to the main menu while the music continues playing in the background. For example, you can get some music going and then switch to another app like Amazon Photos and view your photos with the music playing in the background.

While the music is playing in the background, you can continue to use the playback controls on your remote to control the music. When you press one of the media control buttons, a mini version of "Now Playing" will be displayed on the screen showing the current song and the playback progress.

If you are browsing elsewhere, to quickly go to the currently playing song, press the **Menu** button on the remote. Press the **Back** button to return to your previous browsing location.

To stop the music from playing in the background at any point, press the **Play/Pause** button on your remote.

Viewing Song Lyrics

The Amazon Music app displays the lyrics for most songs on the Now Playing screen. It highlights the lyrics line-by-line as the song plays, enabling you to follow along.

If you are listening to a song that you have uploaded to your library, you can actually use **Down** and **Up** on your remote's navigation trackpad to scroll through the lyrics and jump to a particular part of the song by pressing **Select**. This option is not available for the songs included with Amazon Music because those songs are not saved to your library but streamed from the Amazon Music catalog.

Lyrics cannot be turned off, but you can leave the "Now Playing" screen by pressing **Back** on your remote while the song continues to play in the background.

Note that lyrics are only available for some songs, and lyrics may change as they are provided by an external service. If lyrics are no longer available for a particular song, they will no longer be displayed.

Chapter 12: Bonus: Find Videos to Watch from External Sites

In this chapter, we will cover several external methods that you can use to find content to watch on your Fire TV Stick. The Fire TV already has a very versatile search function, especially with the Alexa feature. However, you can take it a step further and find those hidden gems by using websites with a comprehensive database of movies and TV shows available for online streaming. It can often be easier to find these hidden gems using external websites because you can search by viewer ratings as well as other filters that enable you to refine your search.

Finding Videos at Amazon Video

One way to carry out searches with filters is to do it on the Amazon Video website on your PC. Then you simply add the videos you've found to your **Watchlist.** Whatever you add to your Watchlist on the website will be available on your Watchlist on your Fire TV Stick.

For example, let's say we are looking for historical movies. This is how you would carry out the search:

Go to **www.amazon.com**, using your Internet browser.

On the top left of the window, click **All** > **Movies and Television** > **Prime Video**.

This opens the Prime Video web app.

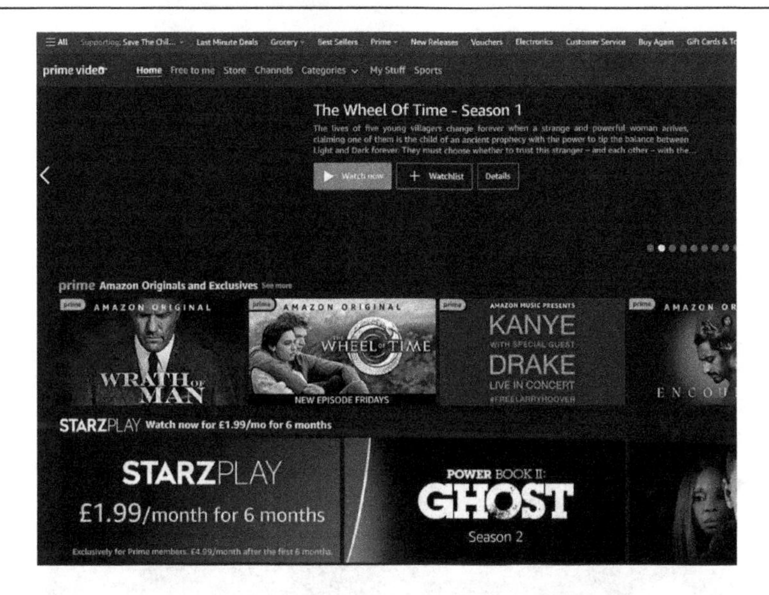

On the menu, click **Categories** to display various categories and genres.

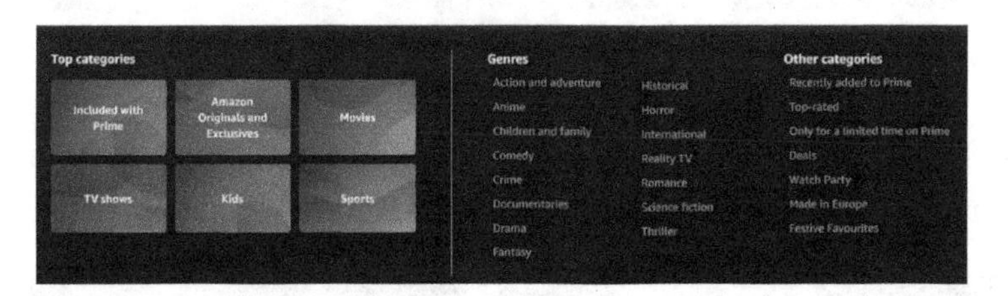

Under Genres, select Historical.

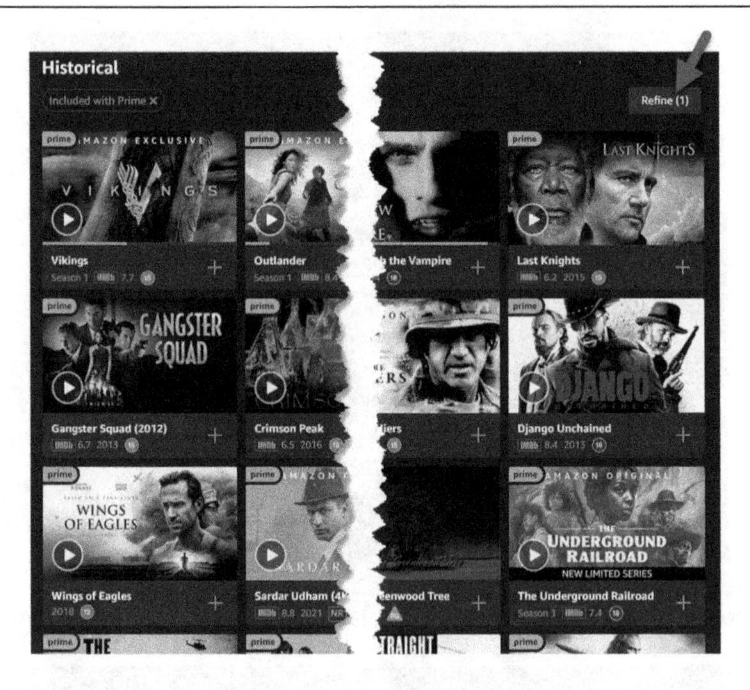

Use the **Refine** button to refine your search results.

Once you identify a video you would like to watch on Fire TV, click the video to go to its details page, and click the **Add to Watchlist** button (+).

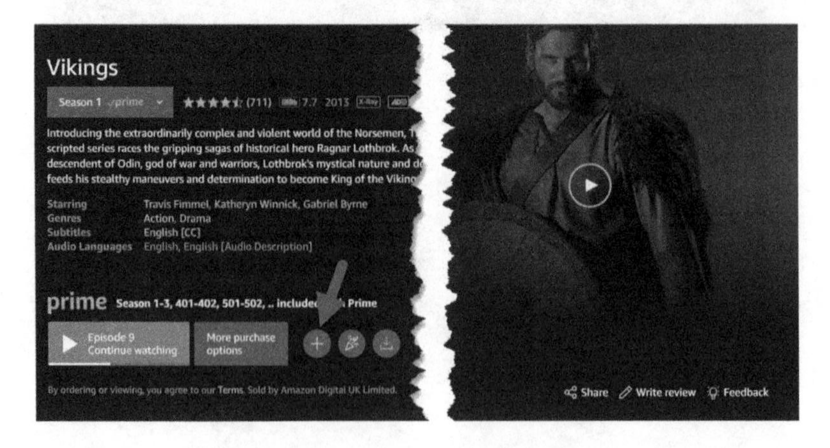

The video will now be available on your Fire TV Watchlist for easy access.

Finding Videos at IMDb

IMDb Ratings

One method of checking if a movie or TV show is worth a try is to look at the IMDb (Internet Movie Database) rating. The IMDb ratings are a good measure because thousands of people carry out the votes. The more votes, the more you tend to have an accurate reflection of the general quality of the movie or TV show.

On Fire TV, the IMDb rating is usually displayed on the details page of a video. You can also use the number of stars from Amazon customer reviews to determine the general quality of a movie or TV show, but the IMDb rating has been known to offer a more accurate reflection. Of course, how good a movie is can be subjective, depending on the viewer's interest, but the IMDb rating offers a general guide.

Using the IMDb Website to Find Movies and TV Shows

IMDb is a once independent (now an Amazon subsidiary) website that has a database of all movies and TV shows, including ratings by viewers from across the world.

Link: http://www.imdb.com/

IMDb is a great resource to find and filter movies to watch based on viewer ratings.

With this website, you can search for highly-rated movies and TV shows that are available to you on Amazon Video or Amazon Prime. The extensive built-in **Advanced Search** feature on the website enables you to search using a wide array of categories and filters.

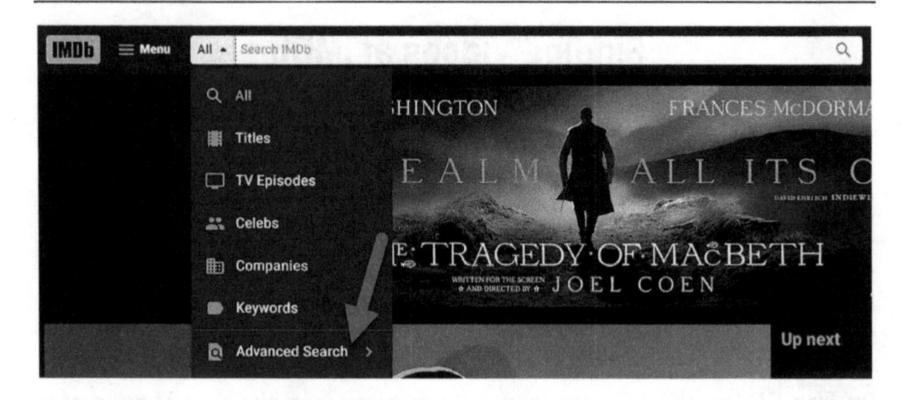

When you click **Advanced Search**, you are taken to the Advanced Search page.

http://www.imdb.com/search/

On this page, under **Movies, TV & Video Games**, click **Advanced Title Search.** This will take you to the Advanced Title Search page.

http://www.imdb.com/search/title

On this page, you have several criteria that you can use to narrow down your search.

You have movie title, title type, release date, user rating, number of votes, genres, title groups, title data, companies, instant watch options, US box office gross, US certificates, color info, countries, keywords, languages, filming locations, popularity, plots, production status, cast/crew, runtime, sound mix, your ratings, and in theatres.

As you can see, it provides an extensive list of filter options that enable you to narrow down the results to find what you want to watch.

Using the Filters

Let's say we are looking for TV shows from 2013 with a rating of 8.5 and above, with at least 5,000 votes (the more votes, the more accurate the ratings reflect the quality).

The image below shows the filters set on the screen. For the ranges, you just need to enter the starting range figure.

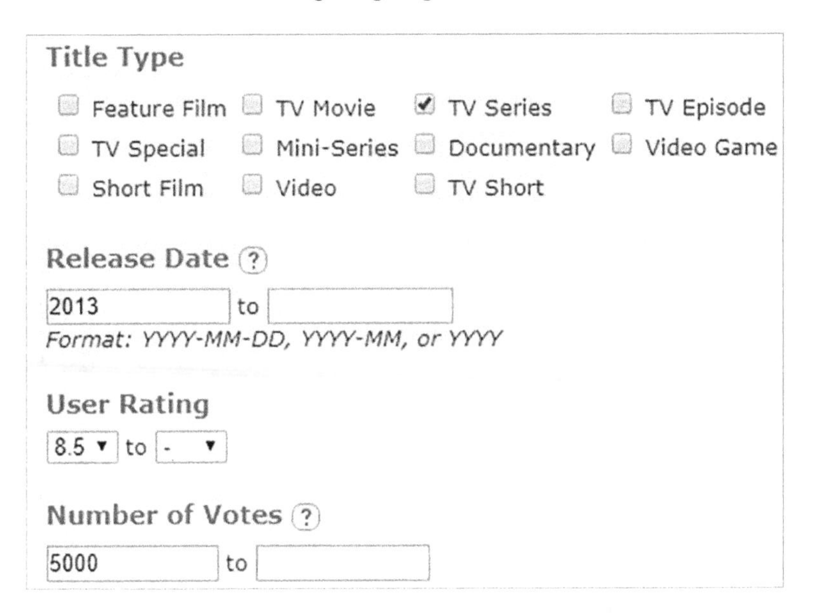

You can scroll down the page and filter your results further with Instant Watch Options.

For instance, if you want to see only videos available for free on your Prime subscription, select US Prime Video (or the equivalent for your country).

Instant Watch Options

- [x] US Prime Video ($0.00 with Membership)
- [ ] US Prime Video (Rent or Buy)
- [ ] UK Prime Video ($0.00 with Membership)
- [ ] UK Prime Video (Rent or Buy)
- [ ] DE Prime Video ($0.00 with Membership)
- [ ] DE Prime Video (Rent or Buy)

When you click on the **Search** button at the bottom of the page, you will get the following results (see image below). Note that the results may vary depending on the date of your search.

You can see that our search returned 17 titles that matched our search criteria.

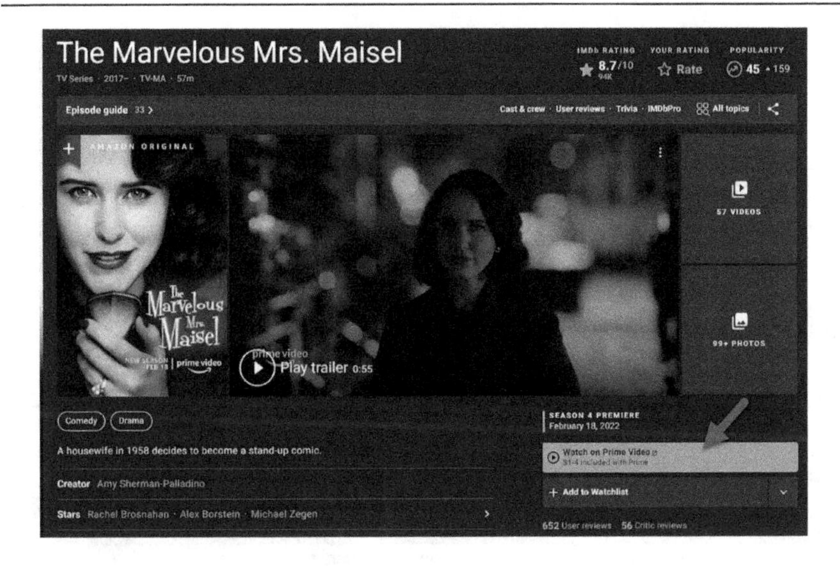

If you go into the details page of one of the search results, you will see that it has a button indicating it is available on Prime Video, and you can watch it with your Prime subscription.

When you click on **Watch on Prime Video**, it'll take you to the details page of the video on the Amazon Video website.

You can now add this to your **Watchlist** for later viewing on your Fire TV Stick.

Finding Content with Rotten Tomatoes

Rotten Tomatoes is a review aggregation website for movies and TV shows. Just like IMDb, it is a renowned website for providing reliable ratings. It also has a comprehensive search feature that enables you to narrow down your search with various search criteria.

Link: http://www.rottentomatoes.com/

To find a movie, for example, on the home page, click on **MOVIES**, then click **Amazon and Amazon Prime.**

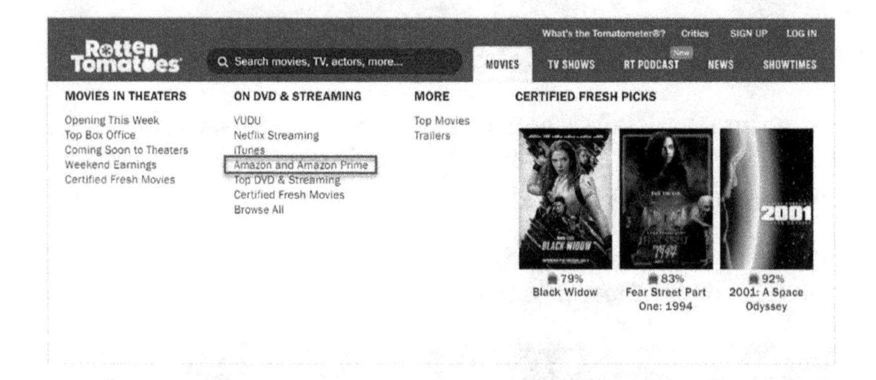

On the next screen, you will see three filter options that allow you to refine your search. The first filter option allows you to filter by the various movie networks. For example, you can select only movies on Amazon Prime if you want.

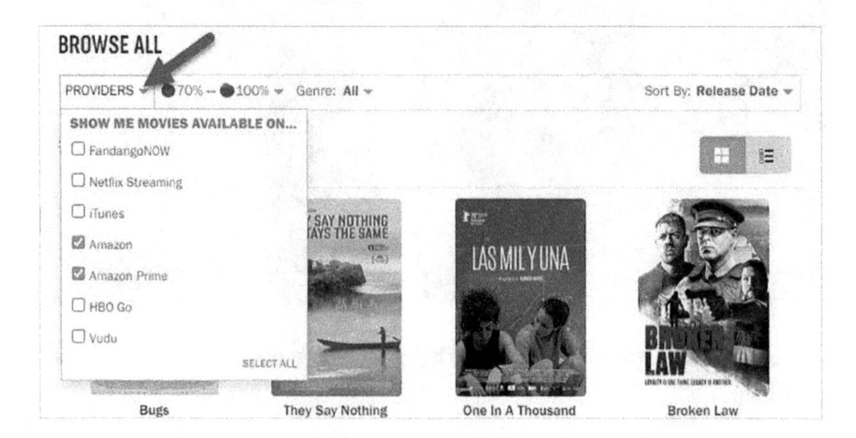

The next filter is called the TOMATOMETER and allows you to filter your selection based on the Rotten Tomato movie rating using a cute sliding scale that ranges from 0% to 100%. For instance, if you only want to display movies rated 70% and above on the Rotten Tomatoes scale, you need to move the sliding scale to the right until 70% is displayed.

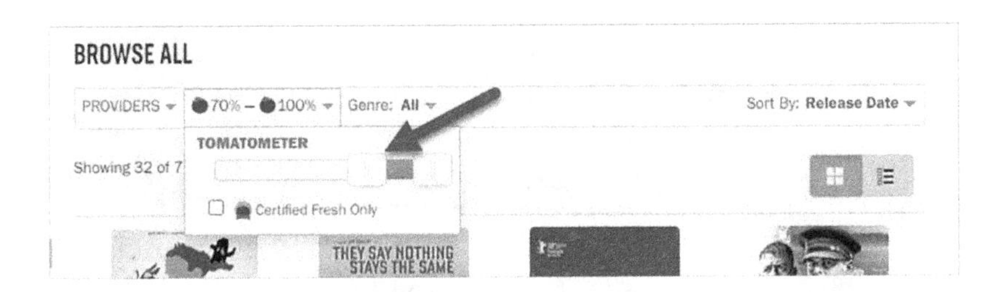

The third filter option is the Genre. You have the option to narrow down your selection to a particular genre or a selection of genres.

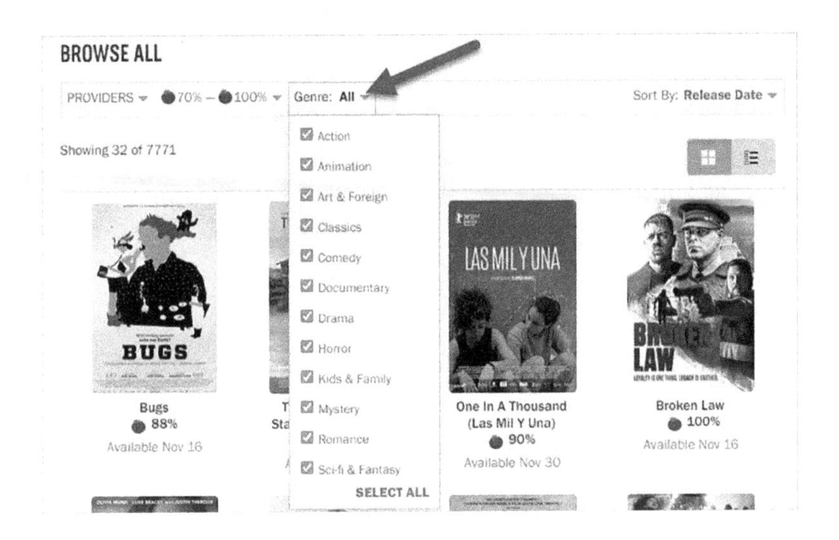

When you select a movie from your results, you will go to the details page of the movie.

When you scroll down the page, you'll see a button that gives you an option to go directly to the movie's page on Amazon Video.

Click on the Prime Video button, and it will take you to the details page of the movie on Amazon's website.

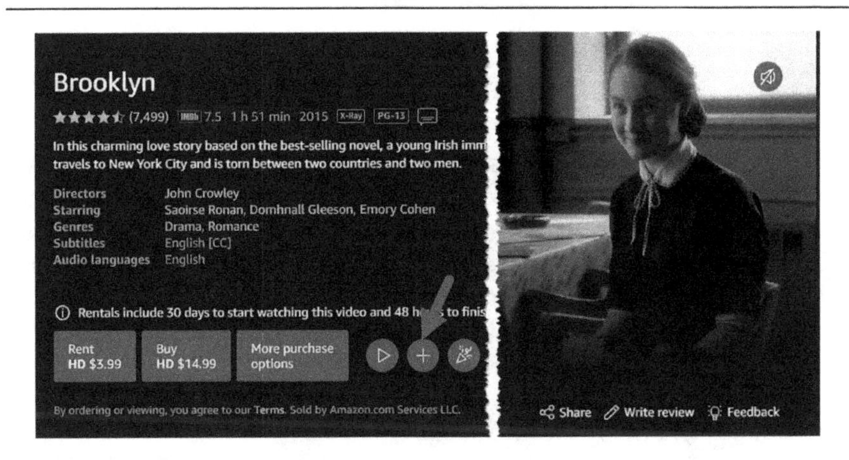

Click the Add to Watchlist button (+) to add it to your **Watchlist** for easy access on your Fire TV Stick.

Finding Content with JustWatch

JustWatch is a website with an extensive database of movies and TV shows from popular streaming networks. You can find movies and TV shows using several search tools to refine your search and narrow down your results. This website is particularly useful for finding those hidden gems, i.e., highly rated movies that are not that popular. Once you find something you want to watch, you can click on a link that takes you directly to the details page of the streaming network.

Website: https://www.justwatch.com/

When you first launch JustWatch, you will be presented with a screen to choose your country of residence. This is important because the selection of videos and the available streaming networks (like Netflix and Amazon Video etc.) will be different depending on your region. You only need to select this once, and the site will automatically default to that country on subsequent use.

The examples here are based on the US site, but you should select the one for your home Amazon account.

The home page gives you several options to filter your search. You can choose from several TV networks, including Amazon, Netflix, Hulu, Apple, BBC iPlayer, etc.

Then you can use the accompanying filters to narrow your search:

- Movies
- TV Shows
- Release year
- Genres
- Price
- IMDb Rating

Click on one of the filters and select your search criteria. For example, if you click **Release year,** you will get a slider that allows you to select a range for the release year of the movie.

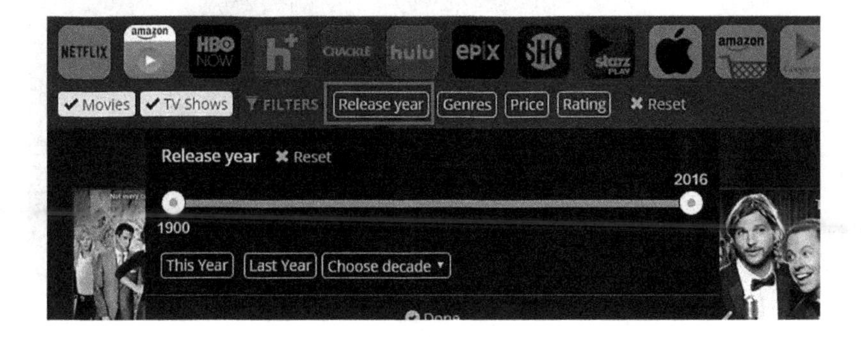

You can click on **Genre** to narrow down your results to specific genres.

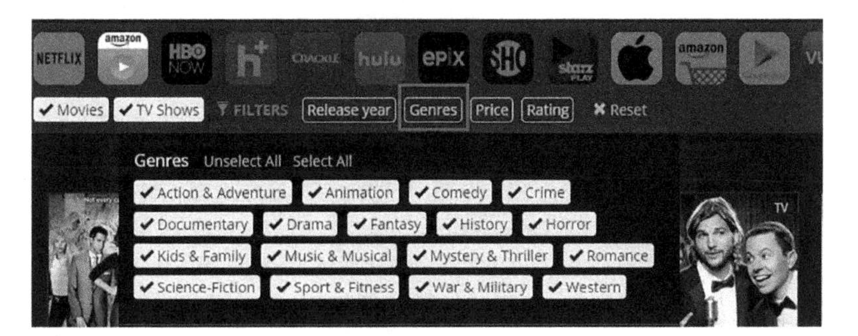

You can also narrow down your search results using the **Rating** filter, which uses ratings from IMDb and Rotten Tomatoes. You select your rating by moving the slider.

Once you've decided on a movie, click on the image to go to the details page.

Here you can click on the Amazon button to go to the video's details page on Amazon Video.

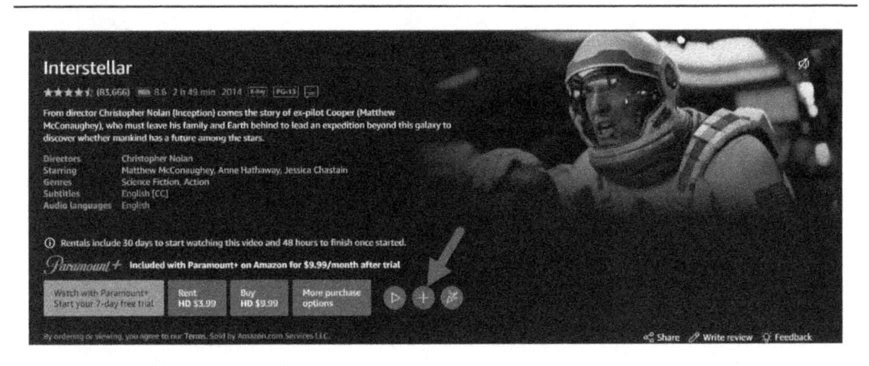

Click the **Add to Watchlist** button (+) to add it to your **Watchlist** for easy access from your Fire TV Stick.

With these three sites, you get an additional set of resources to help you find movies and TV shows to watch on your Fire TV Stick

Afterword: Getting More Help with Fire TV

Thank you for buying this guide. I hope this guide has helped you in setting up your device and helped to improve your viewing experience with your Fire TV.

The Fire TV is a constantly evolving platform, with regular updates to the interface. This guide has focused mainly on the core features of the software that will remain the same despite cosmetic surface changes. This ensures that this book will remain relevant even with later revisions of the Fire TV software.

As mentioned in chapter 1, there are several useful help videos available on Fire TV. You can access them by navigating to:

Setting > Help > Help Videos

The topics currently covered include:

- Welcome to Fire TV
- How to search for content
- Downloading your favorite app
- Personalizing your entertainment experience with profiles
- Fire TV basics
- Setting up your Fire TV
- Creating, editing, and deleting Fire TV profiles
- Mirroring your mobile device on Fire TV
- Customizing the apps on your Fire TV
- Troubleshooting Wi-Fi connection issues

- Pairing your remote using Fire TV Settings or the Fire TV app
- Fixing a blank screen on the Fire TV device

These videos are regularly updated and would come in handy if you want to familiarize yourself with a feature of Fire TV or troubleshoot a problem.

To get more help online with Fire TV, enter "*Amazon Fire TV Help*" in your Internet search browser. This will take you to the appropriate help page for your country.

Alternatively, you can go to the following site, where you'll find quick fixes and troubleshooting steps for different kinds of technical problems in both text and video form:

https://www.amazon.com/gp/help/customer/display.html?nodeId=201729510

About the Author

Nathan George worked in the IT services industry for several years in various roles, including end-user technical support, training, and creating user guides for computer software and peripherals. As an author, he has written several technical books and user guides.

Other Books by Author

Take Your Excel Skills to the Next Level!

Mastering Excel 2019 is your all-in-one guide to Excel 2019. This guide contains everything you need to know to master the basics of Excel and a selection of advanced topics relevant to real-world productivity tasks.

This guide has been designed as a resource for you whether you're an Excel beginner, intermediate user, or power user. You will learn how to use specific features and in what context to use them.

Available at Amazon:
https://www.amazon.com/dp/1916211380

For more books Excel Books visit our website:
https://www.excelbytes.com/excel-books/

Database Creation and Management Made Easy!

This practical guide provides expert advice using a friendly step-by-step approach. You'll learn how to build your own Access databases and enhance them to meet user needs.

Even if you have no formal database training, don't worry. This book starts from the basics and shows you how to structure your data for a relational database using real world examples.

Whether you're new to Access or looking to refresh your skills on this popular database application, **Mastering Access 365** has all the information you need to begin building efficient and robust Access databases today!

This book covers Microsoft Access 2021 and Access for Microsoft 365 (2022 update).

Available at Amazon:
https://www.amazon.com/dp/1916211399

CPSIA information can be obtained
at www.ICGtesting.com
Printed in the USA
LVHW060951010623
748620LV00020B/181